EARTH SCIENCES

Trends in Science

EARTH SCIENCES

Overview by Edward D Young and Margaret Carruthers

Copyright © Helicon Publishing 2001

All rights reserved

Helicon Publishing Ltd
42 Hythe Bridge Street
Oxford OX1 2EP
United Kingdom
E-mail: admin@helicon.co.uk
Web site: http://www.helicon.co.uk

First published 2001

ISBN: 1–85986–368–X

British Library Cataloguing in Publication Data
A catalogue record for this book is available from the
British Library.

Typeset by
Florence Production Ltd, Stoodleigh, Devon
Printed and bound in Great Britain by
Clays Ltd, Bungay, Suffolk

Acknowledgements

Overview
Edward D Young
Margaret Carruthers

Editors

Editorial Director
Hilary McGlynn

Managing Editor
Elena Softley

Editor
Catherine Thompson

Editorial Assistant
Ruth Collier

Content Fulfilment Manager
Tracey Auden

Production Manager
John Normansell

Production Assistant
Stacey Penny

Picture Researcher
Sophie Evans

Contents

Preface *ix*

Part One

Overview 1

Chronology 31

Biographical Sketches 57

Part Two

Directory of Organizations and Institutions 81

Selected Works for Further Reading 91

Web Sites 99

Glossary 123

Appendix *229*

Index *327*

Preface

Scientists engaged in the study of the Earth and the other rocky bodies of the Solar System are troubled by what can only be described as a crisis of identity. The crisis is deep. What was a geologist 20 years ago is now an earth scientist. But is a chemist studying the oceans an earth scientist? Academic departments worldwide are vexed. What are they to call themselves? Should they be known as the Department of Geology, the Department of Earth Sciences, the Department of Earth and Environmental Sciences, the Department of Geological and Geophysical Sciences, or the Department of Earth and Planetary Sciences? In some quarters it is even suggested that the activities of scientists studying the Earth can no longer be described as belonging to a single discipline, and that just as it is rare to find the life sciences under a single roof in most universities today, so too will go the earth sciences.

So this book does not set out to be a definitive account of the state of earth sciences. Rather it is a collection of resources, which will give the reader the necessary grounding to pursue his or her own studies in earth sciences into the next century.

The **chronology** is a year-by-year account of important advances since 1900. Here you will learn how earth sciences has unfolded since the beginning of the last century and be able to trace the many strands of discovery as they unwind throughout the century.

In science it has been customary to associate advances in knowledge with individual scientists – one thinks of Wegener, Einstein, and Darwin – so in the **biography** section you can read about some of the geologists, meteorologists, oceanographers, and other earth scientists who have contributed in some important way to their discipline in the 20th century. Yet in earth sciences, as in other disciplines, the luminaries seem to shine less intensely towards the end of the century than at the beginning. Earth sciences is now "big science", almost wholly reliant on state funding, and the days when a determined individual could dominate a field are almost gone – there are no more Wegeners.

Parallel trends are at work in the **organizations and institutions** of earth sciences. In 1900 a mere handful of institutions were doing serious research in earth sciences, and they were essentially all in Europe and North America. Today, such is the proliferation of research, that we have have been hard pressed to list the institutions that one might regard as key.

One obstacle to the public appreciation of any science is the specialized language, which is why we have included a **glossary** of definitions of

Preface

hundreds of terms from the life sciences, from abiotic factor to zoology, that will inform and enrich your reading.

But where should you start your study of earth sciences? To orient the reader, the book begins with an overview of the field since 1900. Here are the big ideas of the 20th century, and the seed corn for the 21st.

Part One

Overview	1
Chronology	31
Biographical Sketches	57

1 Overview

Introduction

Common use of the term 'earth sciences' is quite new. The word 'geology' literally means 'study of the Earth', and is regarded as the study of the materials and processes that form the Earth, and of Earth history. However, the beginning of the 20th century saw a specialization of fields, which led to the breakup of geology as an all-encompassing discipline. Those who study fossils are grouped as palaeontologists; those who apply principles of physics, such as those of gravity and magnetism, to understanding the Earth are called geophysicists; those who study landforms are known as geomorphologists; those who break down rocks and minerals into their elemental components are classified as geochemists. In fact, very few earth scientists today consider themselves to be classical geologists.

The use of the term earth science is a recent attempt at bringing these sub-disciplines of geology back together as one. In a sense, although earth science is not the modern substitute for the modern definition of geology, it is the modern substitute for the original 18th and 19th century word.

When scientists trained in any scientific discipline turn their attention to the study of the Earth, they are engaged in the earth sciences. The field is therefore unusually broad, encompassing studies of Earth's atmosphere, oceans, crust, and the structure and composition of the planet at great depth. It even extends to the study of the other bodies of the Solar System, including the so-called 'terrestrial' planets and the other rocky bodies of the Solar System. Depending upon their subject of study, biologists, chemists, physicists, mathematicians, and materials scientists can all be counted among earth scientists.

The diversity of the earth sciences is a consequence of the scientific revolutions that transformed the 'natural history' of the 18th and 19th centuries to the modern physical and life sciences of the 21st century. These revolutions facilitated the transition from mere description of the Earth to a more rigorous approach to understanding the interactions among our planet's varied constituents and its evolution as a whole. Today, rather than simply describing layered rocks comprising mountain ranges as they might have done in the 19th century, geologists are relating what they have learned about the formation of those mountains to the chemistry of the oceans and atmosphere and to natural changes in global climate. The geologist can thus be called upon to be conversant in atmospheric chemistry and oceanography.

Overview

Earth sciences are in one sense simply branches of physics, chemistry, and biology. Yet the field is distinguished by virtue of the complexity of the target of study, Earth. Our planet and its Solar System environs are extraordinarily complex systems, and the full spectrum of factors that can influence these systems is seldom, if ever, known. There is also the element of time. Like astronomy, earth sciences have an historical aspect that can involve unfathomable time scales. These distinctions can demand the liberal use of inference as a substitute for verification, a state of affairs that makes most scientists uneasy. The challenge of the 20th century was to ensure that the reliance on inference is kept to a minimum. Out of necessity, then, earth scientists have both instigated and made use of new developments in the physical and life sciences.

Arguably, the most significant advance in earth science of the 20th century was the theory of plate tectonics. This global theory brought order to seemingly unrelated observations made over the previous roughly 200 years of scientific study of the Earth. Driven by a combination of the efforts of insightful and dedicated scientists and technological advances born of war, the discovery that the Earth's surface is laterally mobile ranks as one of the great scientific discoveries of the 20th century.

Understanding Earth's materials

With advances in chemistry, physics, and technology, earth scientists of the 20th century were given new theories, techniques, and equipment with which to analyse the materials that make up the Earth's rocks and minerals. The result was a greater understanding of what rocks and minerals are made of on an elemental scale, the detailed structure of minerals on an atomic scale, and, as a result, a greater understanding of the physical conditions under which the minerals (and the rocks they compose) form. These innovations essentially made it possible to understand (or try to understand) geological processes that occur out of direct observation (for example, the formation of rocks at great depth). Even where they do not necessarily provide new types of information, new technologies have made it possible to analyse materials faster and on a much smaller scale, allowing the science in general to move much more quickly. Whereas a mineralogist of the late 19th century, for instance, relied on a tedious barrage of chemical and physical tests to determine the composition of a mineral, a 21st century mineralogist can use an electron microprobe to acquire detailed information about the composition of many minerals in a rock in just a few hours.

Use of thermodynamics

Due largely to the work of US chemist J Willard Gibbs (1839–1903) and Austrian chemist and physicist Ludwig Boltzmann (1844–1906), the end of the 19th century saw the emergence of chemical thermodynamics as a fundamental branch of physical chemistry. With the advent of thermodynamics came the ability to ascertain the conditions under which minerals, the materials comprising rocks, form. Rather than simply describing rocks, scientists now had the theoretical framework with which to assess the physical and chemical significance of the rocks.

According to the theory of thermodynamics, combinations of minerals exist over a certain range of temperatures and pressures. Pressure increases with depth in the Earth because of the weight of the overlying rock. Temperature also increases with depth because of the rise in pressure, the decay of radioactive elements, and the heat left over from accretion and differentiation. As pressure and temperature of a rock are changed slowly by the rock's burial beneath growing mountains, the rock's minerals react with one another to form new minerals. Early in the 20th century scientists recognized that the distribution of minerals exposed at the surface could be used to reconstruct the size and shapes of ancient mountain belts if the pressures and temperatures that cause them to grow were known. Rocks composed of minerals that grew at great pressures of thousands of atmospheres must have been at one time buried deep in the Earth and then exhumed by erosion of the mountains above. In the years before World War II, Norwegian chemist Victor Moritz Goldschmidt (1888–1947) and Finnish chemist and geologist Pentti Eelis Eskola (1883–1964) showed that thermodynamics could be used to relate minerals at Earth's surface to the physical conditions of their growth. Through the efforts of the likes of Goldschmidt and Eskola, it became possible to decipher the evolution of ancient mountains belts.

Improvements within thermodynamics

Earth scientists not only made use of thermodynamics in the 20th century, they improved upon it as well. Using X-rays, mineral chemists could see that replacement of one element by another in a mineral did not alter its structure. The fact that their chemical compositions could be changed without affecting their structure meant that minerals had to be treated differently from the solutions that most chemists were accustomed to working with. Geoscientists adapted thermodynamics so that it could be used with these peculiar constraints imposed by crystalline substances.

Similarly, Russian chemist D S Korzhinskii (1899–1985) showed that traditional thermodynamics could not be used to predict the minerals that will form in a rock when water flows through its pores. Since water commonly flows through rocks, even at considerable depth, in 1959 Korzhinskii devised ways of expanding the rules of thermodynamics to allow for reaction between rock and flowing fluids.

The industrial revolution had afforded the technology to verify the concepts of thermodynamics in the laboratory. In the earth sciences this process began in earnest in 1905 when the Carnegie Institution of Washington, DC, established the Geophysical Laboratory staffed with physicists, chemists, and geologists. It was here that Canadian geologist Norman Levi Bowen (1887–1956) and his colleagues revolutionized geology by establishing the chemical and physical state of molten and solid rock at high temperatures in the laboratory. Bowen's experiments allowed him to suggest that Earth's crust was the product of the melting of parts of the mantle, a process known as differentiation. His book *The Evolution of the Igneous Rocks* (1928) is considered the most influential work on the origin of igneous rocks of the 20th century, and firmly established the potential of the physical chemical approach to geology.

Minerals at an atomic level

The properties of minerals are the result of the different ways in which their constituent atoms are bonded together chemically. Understanding the behaviour of rocks at high pressures and temperatures in the Earth requires an understanding of materials at the atomic level. Early in the 20th century, US chemist, biologist, and twice Nobel laureate Linus Pauling (1901–1994) formulated the electronegativity scale and so provided a logical means for describing how atoms bond with one another. Pauling's first paper, in 1923, dealt with the crystal chemistry of the mineral molybdenite. The first diagram in his classical monograph *The Nature of the Chemical Bond* (1939) shows the crystalline structure of rock salt. In *The Nature of the Chemical Bond*, Pauling enumerated a set of rules for rationalizing the structures of crystalline substances (such as minerals). Pauling's rules, as they have come to be known, are still used as a first-order means for understanding mineral structures and his work ranks as one of the most important advances in understanding rock materials.

It was Pauling who elucidated the nature of the bond between silicon and oxygen, the fundamental building blocks of Earth and the other rocky planets. In 1980 when his description of the Si-O bond was questioned on the basis of a more modern approach, erstwhile mineral chemist Pauling took the time to publish an article in the *American Mineralogist* to restate his case.

The early work on bonding laid the foundations for the modern discipline of computational mineral physics. Today mineral physicists use supercomputers to simulate forces on individual atoms and bonding at the scale of electrons. With such calculations they can predict the behaviour of minerals at extreme conditions inaccessible to humans, including the deep interior of the Earth.

An understanding of bonding is also useful for studying the chemical processes that occur at the interfaces between Earth's regolith, atmosphere, and hydrosphere. As a result, new subdisciplines such as environmental surface chemistry have evolved. Mathematical models for atomic and molecular bonding are used to predict the rates at which atoms in the environment adsorb (stick) and desorb (unstick) to mineral surfaces. Studies of this kind can be applied to better understand such things as the rates of stream acidification and groundwater contamination and to devise methods for neutralizing contaminants. Development of new surface-sensitive analytical techniques like atomic force microscopy (AFM) are aiding scientists in the study of water–mineral interactions.

Earth's deep interior

At the beginning of the 21st century, drilling deep into the Earth is still the realm of science fiction. Engineering and financial constraints have limited the deepest holes to about 13 km/8 mi. As a consequence, direct examination of our planet is confined to the outer shell, which comprises less than 1% of its volume. Geophysics provides us with a 'window' into the deep interior of Earth and that of other bodies of the Solar System. Studies of Earth's magnetism, gravity, and seismicity, the traditional subdisciplines of geophysics, were all blossoming in the early part of the last century.

Devices for measuring Earth's magnetic field were being towed across land and oceans around the world by 1905. As early as 1906 two distinct forms of rock magnetism had been discovered. Some rocks were magnetized with their north 'poles' parallel to Earth's present magnetic field while others were magnetized with their poles reversed with respect to the Earth. In the 1960s this observation would prove to be crucial in the revolutionary discoveries of seafloor spreading and plate tectonics.

Geophysics as a tool for petroleum exploration began in 1922. By that time the association between salt domes and oil in the Gulf of Mexico was well known. Since salt is much less dense than rock, the gravitational field above salt domes should be distorted. The torsion balance, a device for measuring gravitational field strength, was used to locate salt domes, and thus oil, in the Gulf. For the first time, geophysics had been used for prospecting.

Use of seismic waves

At about this same time, artificial seismic waves were being tested as means for exploring the Earth's shallow structures with an eye toward finding more oil, and by 1930 seismic reflection was well established as the most widely used geophysical tool for petroleum exploration.

In the early part of the 20th century, Irish seismologist Richard Oldham (1858–1936) showed that seismometers recorded seismic waves from earthquakes all over the world. He recognized that because the speed that the waves travel depends on the density of the layers they travel through, seismic waves could be used to unravel the layered structure of the Earth.

In 1906, Oldham calculated that seismic waves slow down as they pass through the centre of the Earth. This suggested that below the mantle was a central core to the Earth, with different properties from the mantle. Furthermore, seismologists noticed that the P waves (compressional waves) were travelling abnormally slowly through this core and that the S waves (transverse waves) did not seem to be travelling through this central core at all. Oldham showed that the culprit blocking the S waves and slowing down the P waves was a core that behaved like a fluid – S waves cannot travel through fluids. In 1914 German-born US seismologist Beno Gutenberg calculated that the surface of the newly discovered core lay at a depth of approximately 2,900 km/1,800 mi.

Two decades later, in 1936, Danish seismologist Inge Lehmann (1888–1993) noticed that the seismic signals from the core were much more complicated, and pointed towards the existence of a denser inner core within the liquid outer core. Further studies suggested that the inner core was most likely solid, and in 1970s, seismologists finally detected reflections off the inner core, which essentially proved its existence.

By the middle of the 20th century the velocities of seismic waves revealed even greater detail about the layered structure of our planet. The speed with which a seismic wave moves through rock depends upon the density of the rock. The greater the density, the greater the velocity. It was known that there were several planet-wide changes in seismic velocity with depth. The shallowest, the Mohorovicic discontinuity, named in honour of Croatian seismologist Andrija Mohorovičić (1857–1936), lies 5–10 km/ 3–6 mi beneath the ocean floor and approximately 35 km/22 mi beneath the surface of the continents. Below the 'Moho' seismic waves travel nearly 20% faster than above and it is regarded as the bottom of Earth's crust.

In the last two decades of the 20th century, three-dimensional imaging of Earth's seismic structure, a technique known as 'global seismic tomography', has provided snapshots of the mantle in much the same way that CAT (computer-aided tomography) scans are used to image the human brain in medicine. Using seismic tomography, geophysicists like A M Dziewonski (1936–) of Harvard University and J H Woodhouse of Oxford University have shown that Earth's tectonic plates descend deep into the mantle, and that large portions of cold dense mantle slowly sink and are replaced by more buoyant warmer mantle. The ability to relate deep mantle structures to surface features of the planet constitutes a major advance in the earth sciences.

The Earth is often described as being composed of an outer thin crust, a mantle, and a core. But as early as the 1920s it was known that there were other dramatic changes in seismic velocities that could not be explained by this simple picture. Velocities change dramatically with depth in the mantle at depths of 100–200 km/60–125 mi, and again at 400 and 600 km (250 and 375 mi). These transitions are every bit as fundamental as the distinction between the core, mantle, and crust, but their nature is the subject of continued research. The change in seismic velocity at about 600 km/375 mi depth is used to define the transition between the upper and lower mantle.

In order to relate seismic velocities, the window into Earth's deep interior, to the unseen materials that make up our planet at depth, earth scientists of the 20th century turned to the emerging field of high-pressure physics. Deep in the Earth, rocks are subjected to crushing pressures due to the burden of overlying rock. At the bottom of the crust, pressures are 20,000 times that exerted by our atmosphere. Physical chemistry suggests that the mineralogy of a rock will change in response to these immense pressures. High-pressure physics has provided the means for assessing what those changes are likely to be.

Hot ice and artificial diamonds

At the beginning of the 20th century the maximum pressure obtainable in the laboratory was of the order of 2,000 atmospheres. In 1910, US physicist Percy Bridgman (1882–1961) invented a device called the collar that allowed him to confine materials between two pistons. With his new device Bridgman was able to squeeze all kinds of materials to pressures comparable to the base of Earth's crust. He discovered, for example, that he could squeeze water to produce a form of ice that could exist at temperatures near the boiling point of ordinary liquid water. This 'hot ice' produced a sensation in the media.

Following World War II, the pursuit of the first synthetic diamonds led to rapid developments in high-pressure technology. Inexpensive diamonds would be invaluable for machining weaponry. While exploring ways of producing high temperatures and pressures, Loring Coes (1915–), a researcher at the Norton Company in the USA, learned how to grow minerals at temperatures and pressures as high as those of the Earth's mantle. Coes could grow beautiful minerals like red garnets, minerals thought to be produced deep in the Earth at high pressures. Geoscientists took note, and soon they too learned how to mimic Earth's pressures and temperatures in the laboratory. Coes discovered a high-pressure form of quartz that is named after him, coesite. Natural coesite was later discovered to be present in Meteor Crater, Arizona, USA, where it had been produced by the great pressures during meteorite impact. His research showed that new abrasives of economic value could be produced at high pressures. In 1954 researchers at General Electric in the USA succeeded in producing the first synthetic diamonds.

Overview

Diamond anvil cell

The US military fostered high-pressure research in other ways too. Investigations into how metals behave at high pressures were useful for designing submarines. The Chicago University Institute of Metals designed a new pressure device made of a single diamond with a hole drilled in it. Diamond was used because it is made of carbon, which is relatively transparent to X-rays. The researchers wanted to pass X-rays through metals while they were being squeezed in order to examine how their structures

Diamond anvil cell used in laboratory experiments to simulate the high pressure and temperature found inside planets. Diamond is the hardest and least compressible material and is also transparent to most of the spectrum of electromagnetic radiation, including γ-rays, X-rays, portions of ultraviolet, visible light, and most of the infrared region. The sample being studied is placed between two diamonds and is contained on the sides by a metal gasket. In this configuration, very little force is required to create extremely large pressures in the sample chamber and, because of the transparency of diamond, the sample may be examined in situ (while at elevated pressure) by, for example, optical microscope or a spectroscope. High Pressure Diamond Optics, Inc (http://www.hpdo.com)

changed. This simple device evolved into the diamond anvil cell (DAC), in which samples are squeezed between two diamonds using a small hand-sized anvil. The principle is a simple one: a small force applied to a tiny area like the head of a cut diamond produces very large pressures.

With the DAC physicists, chemists, and earth scientists would be able to produce pressures exceeding one million atmospheres. Indeed, this milestone was achieved by two US earth scientists, David Mao and Peter Bell of the Carnegie Institution of Washington's Geophysical Laboratory, in 1975.

In 1974 Australian earth scientist John Liu suggested that most of the Earth is probably composed of a mineral with a structure wildly different from the minerals found near the surface. The majority of rocks are made from silicon chemically bound to oxygen. With very few exceptions, tiny silicon atoms surrounded by four larger oxygen atoms are the major component of most minerals that make up familiar rocks such as granite. But Liu's new mineral, magnesium silicate perovskite, is different because the small silicon atoms are surrounded by six oxygens rather than four. This means that magnesium silicate perovskite is very much denser than the minerals familiar to geologists. Liu made his discovery by squeezing minerals thought to have the chemical composition of the mantle with the DAC while heating the mineral with a laser beam (in order to simulate mantle conditions). Amazingly, the pressure at which the transition to perovskite occurred corresponded nearly exactly to the depth of the transition between the upper and lower mantle defined by seismic velocities. Liu's discovery showed that the increase in seismic wave velocities at the transition between the upper and lower mantle was caused by the conversion of 'normal' minerals in which four oxygens surround silicon, to minerals like perovskite, in which six oxygens are tightly packed around each atom of silicon. This work sparked a new era of cooperation between seismologists and high-pressure researchers in the earth sciences.

The success of the DAC has broadened the scope of high-pressure research. High-pressure researchers at institutions like the Geophysical Laboratory and the Bavarian Research Institute of Experimental Geochemistry and Geophysics (Bayerisches Geoinstitut) in Germany are going beyond conventional boundaries of the geological sciences to study the interiors of other planets in our Solar System.

The giant planets like Jupiter and Saturn are composed primarily of hydrogen, the most abundant gas of interstellar space. What must these planets be like deep in their interiors, where pressures exceed one million atmospheres? In 1988 it was reported that a solid form of hydrogen had been obtained by compressing hydrogen gas to pressures of between one and two million atmospheres. This solid appeared to be metallic. Although the validity of the particular claim is debated, the implication is clear; the interiors of giant planets may be composed of metallic hydrogen.

Core temperature

The melting temperature of iron at very high pressures is another example of fundamental research into the behaviour of materials with implications for planetary interiors. Surprisingly, we still do not know the temperature of Earth's core. Since the core is thought to be mostly made of iron, and the outer core is liquid, the melting temperature of iron would prove invaluable for estimating core temperature. At first, determining the temperature at which iron melts seems trivial. After all, molten iron has been used to produce steel since before the turn of the 20th century. Surely, one might think, all that is necessary is an accurate measurement of a batch of molten iron in some foundry. But melting temperatures change with pressure, and so they must be measured at core pressures in order to be of use. This is no easy task and at present there are conflicting results obtained with the DAC.

Obtaining accurate estimates of the temperature of the core is a prerequisite for understanding the physics of the planet as a whole. Earth, unlike Mars which ran out of heat long ago, is still geologically active because of the heat contained within. If we are to understand what causes the tectonic plates to move, for example, we must have a better understanding of the temperature of the core.

Geological time

In the year 1900 the age of the Earth was estimated by comparing the amount of salt in the oceans to the rate at which salt was being delivered to the seas by rivers. The calculation suggested that Earth was approximately 90 million years old. Discoveries of radioactivity and isotopes changed forever our view of geological time.

By 1917, due largely to the work of New Zealand-born British physicist Ernest Rutherford (1871–1937) a coherent picture of the atom was emerging. The mass of the atom lay in its nucleus and its chemical behaviour was a function of the tiny electrically charged electrons that surround the nucleus. The number of neutral particles in the nucleus, called neutrons, was found to be variable, giving rise to different masses (or weights) of the same element. The different masses of a given element were dubbed isotopes. Radioactivity, discovered at the end of the 19th century, was rightly identified as the result of decay of an isotope of one element to form another element.

In 1905, US chemist Bertram Boltwood (1870–1927), working with Rutherford, suggested that some isotopes of lead were the product of the decay of some isotopes in uranium, and in 1907 Boltwood set out to show that the relative numbers of parent uranium isotopes and the lead isotopes they produce could be used to determine the ages of uranium-bearing minerals. Using lead-to-uranium ratios, Boltwood showed that ages of rocks from several settings varied 410–2,200 million years. These dates indicated an antiquity not previously considered. The science of isotope dating was born.

Numerous isotope decay series have been used to date rocks since the time of Boltwood, including the decay of rubidium isotopes to produce strontium, and the decay of potassium to produce argon. Over the past two decades, new methods like heating tiny mineral grains with lasers to release argon have drastically reduced the amount of material required for dating. The solid Earth is now known to be approximately 4,500 million years old, and uranium–lead dating of meteorites has allowed geoscientists to conclude that the Solar System began to form from an interstellar cloud of dust and gas some 4,560 million years before present.

Chemistry of Earth's near surface

The term isotope invariably conjures up the spectre of radioactivity to nonscientists, but most chemical elements are composed of several isotopes. The majority of these isotopes, especially for the lighter elements, are not radioactive. The relative abundances of the so-called stable isotopes have proven to be invaluable in the modern earth sciences as means for following the movements of the chemical elements back and forth between Earth's atmosphere, hydrosphere, biosphere, rocky crust, and regolith.

In 1931, US chemist Harold C Urey (1893–1981) discovered deuterium, a heavy isotope of hydrogen, ushering in the modern field of stable isotope geochemistry. Urey and colleagues later succeeded in isolating the rare heavy isotopes of several elements, including oxygen, carbon, nitrogen, and sulphur. Although chemically similar, isotopes of these lighter elements were found to be separable on the basis of mass alone. For example, water molecules composed of heavy isotopes of oxygen are less likely to boil away from a body of water by evaporation than are water molecules composed of the more abundant light isotope of oxygen. Thus rain water, having been derived from evaporation of the oceans, has on average more of the light isotope of oxygen.

Because they can be separated by processes like evaporation, condensation, and chemical reaction, the stable isotopes of an element are like labels that can be used to trace elements as they pass back and forth from one part of the Earth to another. Moreover, quantum mechanics showed that the amount of the rare heavy isotopes of elements like carbon, oxygen, and nitrogen relative to the amount of their more abundant light isotopes change in predictable ways with temperature. On this basis Urey wrote a seminal paper in 1948 in which he foretold the use of stable isotopes as 'geothermometers' and chemical tracers in the earth sciences over the next 50 years.

General Electric researchers use a mass spectrometer to isolate uranium in the 1940s.
Corbis/Schenectady Museum; Electrical History Foundation

The abundance of isotopes of oxygen and carbon in the hard parts of marine organisms past and present are used routinely in the earth sciences to deduce temperatures in the oceans, and these same isotopes are being used to reconstruct the diets, physiology, and ecology of long extinct creatures. Measurements of stable isotopes in teeth and bones give clues as to whether or not mammoths were hunted to extinction by humans or died out because of changes in global climate. Isotopes may even tell us if dinosaurs were warm-blooded or cold-blooded. Indeed, an entirely new field, known as biogeochemistry, has evolved around the ability to use stable isotopes as means for monitoring how chemical elements pass back and forth between living organisms and Earth's hydrosphere, atmosphere, and lithosphere.

Since the time of Urey it had been thought that the separation of two isotopes was always in proportion to the difference in molecular weight. In the early 1980s US chemist Mark Thiemens (1950–) and his colleagues performed experiments in the laboratory that proved otherwise. They demonstrated that the production of ozone by the joining together of three atoms of oxygen in the upper atmosphere causes separation of the two different heavy isotopes of oxygen relative to the one lightest isotope *independent* of their masses.

Thiemens immediately suggested that other reactions might behave similarly, and that these so-called symmetry induced isotope effects could explain the strange abundance of oxygen isotopes seen in some very old objects within stony meteorites. All at once Thiemens's discovery grabbed the attention of geochemists, cosmochemists, and atmospheric chemists. Today symmetry induced isotope effects are being used to track mixing in Earth's atmosphere. Analysis of Martian meteorites suggests that similar reactions may take place on Mars. Debate surrounding the significance of anomalous abundances of oxygen isotopes in meteorites continues.

Technological advances like refinement of accelerator mass spectrometer methods and laser sampling allow earth scientists to make use of more and more isotopes. Cosmogenic nuclides are isotopes produced as a consequence of bombardment by cosmic rays. Cosmogenic nuclides that decay rapidly, like the radioactive form of beryllium, are being used with increasing frequency to date recent fault movements and to trace past ocean circulation patterns.

Climate and global warming

At the start of the 20th century, Swedish chemist and Nobel laureate Svante Arrhenius (1859–1927) suggested that changes in the amount of carbon dioxide gas in the air have influenced Earth's climate over time. The chemical bonds between carbon and oxygen in carbon dioxide cause these gas molecules to vibrate at low frequencies corresponding to infrared radiation, or radiant heat. The result is that heat from Earth's surfaces, warmed by high-frequency ultraviolet light from the Sun, is absorbed by the vibrations within carbon dioxide molecules in the air and so cannot escape. The trapping of heat by atmospheric gases is referred to as the greenhouse effect.

Arrhenius' greenhouse effect has become a subject of intense study in the earth sciences in recent decades because of an increase in average global temperature of approximately 0.5°C/1°F over the past century that has come to be known as global warming. Recent melting and collapse of the Larsen Ice Shelf of Antarctica is one obvious consequence of global warming. By studying the geological and historical records, earth scientists have established that global temperature has been highly variable in Earth history and that fluctuations in global temperature have occurred even in historical times. The recent rise in temperature is therefore not in itself unusual. However, it happens to coincide with the spread of industrialization, giving rise to the hypothesis that it has resulted from a human-generated greenhouse effect in which Earth's radiant heat is trapped by atmospheric pollutants, especially carbon dioxide gas. At issue is to what extent global warming is anthropogenic, resulting from human-made gases in the atmosphere, or simply a natural fluctuation like so many others in Earth history.

Earth scientists must consider, for example, that the present episode of global warming has thus far still left England approximately 1° C cooler than during the peak of the so-called Medieval Warm Period (AD 1000–1400). The latter was part of a purely natural climatic fluctuation on a global scale. With respect to historical times, the interval between the Medieval Warm Period and the rise in temperatures we see today was unusually cold throughout the world. Some scientists argue on this basis that, rather than being an anomaly, the present warming is a return to a more 'normal' state. Assessing the impact of humankind on global climate is complicated by the natural variability on both geological and human time scales. There is no question however, that there has been a steady rise in atmospheric carbon dioxide levels since the onset of the industrial revolution, that carbon dioxide is the main product of burning fossil fuels such as coal, oil, and gas (carbon + oxygen = carbon dioxide), and that carbon dioxide is a major greenhouse gas.

Fluctuations of CO_2

In order to evaluate the likely influences of recent changes in the amounts of greenhouse gases in our atmosphere, earth scientists require a record of fluctuations in atmospheric carbon dioxide in the past. Tiny samples of air trapped in polar ice provide one such record for the more recent historical past. The ice preserves annual layering that can be counted like tree rings to obtain the age of each layer. Trapped bubbles of air thus provide a record of how the chemistry of the atmosphere has evolved over the lifetime of the ice.

At the beginning of the 21st century, improved understanding of the geochemistry of carbon is providing the means for correlating carbon dioxide in air to climate change over geological time scales. Assessing the role of carbon dioxide as a cause for past fluctuations in climate requires a thorough understanding of how carbon moves in and out of the atmosphere. It is still unknown, for example, to what extent the oceans can mitigate increases in atmospheric carbon dioxide. Because the processes involved are slow in human terms, one of the most effective ways of understanding the slow exchange of carbon between Earth's atmosphere, hydrosphere, biosphere, and lithosphere is through examination of the chemistry of carbon as recorded in the geology of the planet. Results of such studies in the last decade have shown that the amount of carbon dioxide in the atmosphere today is unusually low in comparison to most of the rest of the last 600 million years. The onset of the ice ages, times of advance of ice sheets across the continents that began approximately 1 million years ago, has been attributed to the small amount of carbon dioxide in the atmosphere and the average cooling that should result. The link between atmospheric carbon dioxide concentration and ice ages is bolstered by these same studies showing that the only other time of significant continental glaciation, about 300 million years ago, was also a time when the amount of carbon dioxide in the atmosphere was anomalously low, like today.

The connection between short-lived changes in climate, of the order of a single year, and catastrophic geological phenomena like volcanic eruptions was known as far back as the 18th century. In the last two decades it has become clear that longer-term changes in global climate are influenced by fundamental processes like mountain building and mineral reactions at great depths in the Earth. Today, for example, it is suggested that the single most influential factor in global climate over the past 50 million years has been the rise of the Tibetan Plateau.

Changes in Earth's orbit

Long-term changes in climate can be related to changes in Earth's orbit relative to the Sun. In the early part of the century it was known that wobble of Earth's rotation axis occurs with a periodicity of 22,000 years, that variations in axis tilt repeat every 40,000 years, and that ellipticity, or eccentricity, of the orbit changes with a periodicity of about 100,000 years. In 1920 Yugoslavian meteorologist and mathematician Milutin Milankovitch (1879–1959) showed that the amount of energy, or heat, received by Earth from the Sun varies with these long-term changes in orbit. Changes in solar radiation with orbit are now known as Milankovitch cycles, and their capacity for altering Earth's climate has been the subject of study ever since their discovery. Over the next 20 years Milankovitch and others promulgated the notion that advance of ice sheets during the ice ages and their subsequent retreats might be correlated with Milankovitch cycles. Special emphasis was placed on correspondence between 100,000-year cyclicity in the ice ages and orbital eccentricity. Decades later, using the isotopes of oxygen in remains of marine organisms as 'palaeothermometers', scientists correlated changes in ocean temperatures over hundreds of thousands of years to Milankovitch cycles. Periodic changes in Earth's orbit due to interactions among bodies of the Solar System is still not fully understood, and the influence of 'orbital forcing' on cyclical climate change remains an important area of research in the earth sciences at the beginning of the 21st century. Most recently, it has been suggested that periodic changes in the amount of interplanetary debris, principally dust, encountered by Earth as it travels around the Sun can affect our climate.

Life on Earth

Studies of the origin and evolution of life are, perhaps above all other scientific pursuits, truly interdisciplinary, requiring synergetic relationships among geologists, biologists, biochemists, and organic and inorganic chemists.

Metazoa, or animals, first appeared in the fossil record approximately 680 million years ago. The abrupt appearance of multicellular animals implied that early life had evolved at an unrealistically rapid pace. Scientists of the late 19th and early 20th centuries were thus prompted to search for evidence of more primitive forms of life in older, apparently barren rocks. The hunt produced some scant fossil evidence for very ancient life, but it was not until the discovery, in 1954, of a plethora of diverse fossil microscopic organisms in the Gunflint rocks along the north shore of Lake Superior in the USA that the existence of pre-metazoan life became widely accepted. The Gunflint biota proved that there had been significant evolutionary activity by at least around 2,000 million years before present. In the decades that have followed, studies of stromatolites, rocks made from ancient sediments laid down on sticky mats composed of cyanobacteria (also called blue-green algae), have shown that life existed at least as far back as 3,600 million years ago.

In the late 1990s, measurements of carbon isotope abundances in graphite from a rock formed nearly 3,800 million years ago are consistent with (although not proof of) the existence of life. The age of the rock is, however, in question, and evaluation of this evidence for ancient life requires reference to the geology of the Solar System as well as the biogeochemistry of Earth. Studies of craters that abound on the Moon and on the rocky surfaces of other bodies of the Solar System show that 3,800 million years ago impacts by asteroid-sized objects – vestiges of planet formation – were commonplace. Life, it is argued, was unlikely to have survived under such hostile conditions.

Plants produce oxygen through photosynthesis. The geological record shows that oxygen was not always abundant in our atmosphere. When did Earth's atmosphere become enriched in oxygen? Was it the result of photosynthesis or the exhaustion of the supply of iron that was being rusted by reaction with oxygen? Answers to these questions require a synthesis of palaeobiological and geological data. For example, in order to evaluate the influence of life on Earth's atmosphere, it is necessary to have some understanding of the atmosphere before there was life. One possibility is that the atmosphere was exhaled from volcanoes. Geochemists thus study rocks from which the primordial atmosphere could have been expelled so that they can determine the amount of oxygen that might have existed prior to photosynthesis.

Meteoriticists ponder the effects that bombardment by asteroids and comets might have had on Earth's early atmosphere. It has been suggested, for example, that the water necessary to form oceans and life itself may have come from the melted ice of comets that struck the Earth more than 4 billion years ago.

Primordial soup

The origin of organic compounds on our planet has yet to be resolved, despite intensive and creative study over the past several decades. In 1953 then graduate student Stanley Miller (1930–), working with US chemist Harold Urey performed an experiment that showed how organic compounds might have begun on Earth. They exposed a warm, gaseous mixture of inorganic compounds, including ammonia, methane, hydrogen, and water, to electrical discharges. The gases were meant to represent the pre-biotic atmosphere, the proverbial primordial soup, and the electrical discharges simulated lightning. After several days Miller and Urey produced amino acids, the ubiquitous building blocks of life that allow chemical reactions to take place in the cells of living things. Yet the Miller–Urey reactions are not the only source of amino acids. They are also abundant in certain kinds of primitive stony meteorites and similar objects undoubtedly struck the Earth with comparative frequency early in its history. Could life have begun as a result of the bombardment by organic-rich asteroids?

Toward the end of the 20th century, the capacity to separate organic compounds by various forms of chromatography permitted studies of fossil life to be extended to the molecular level. As more amino acids were identified in ancient materials, the method of racemization age dating was developed. This technique is based on the fact that amino acids can exist in two forms, one being a mirror image of the other. In living organisms only one form exists. Upon death, the living form transforms spontaneously to its mirror image at a steady rate over time. The amount of the living form relative to the racemized form thus provides an estimate of the time since death. Organic-bearing fossils can be dated as far back as 20 million years with this method.

The geochemistry of organic compounds has provided new insights into the fate of biota after death. US biogeochemist T C Hoering showed that carbon isotopes of organic compounds extracted from ancient rocks could be used to learn about the processes that give rise to petroleum.

The phrase 'you are what you eat' applies to the isotopes of both modern and ancient organisms. Increasingly, biogeochemists are able to reconstruct the food chains of creatures ranging in complexity from bacteria to elephants and even humans. Comparisons of the isotope effects of metabolic processes in modern and ancient biota is an evolving tool in the earth sciences. For example, it has been shown that the relative abundances of the stable isotopes of carbon from modern blue-green algae that thrive in hot springs rich in carbon dioxide are similar to those preserved in ancient, Precambrian stromatolites. The implication is that Earth's atmosphere was richer in carbon dioxide in the Precambrian than it is today. Such conclusions would not have been possible prior to the confluence of technology and expertise from a variety of scientific specialities that is, for the lack of a better term, a subdiscipline of biogeochemistry.

Mass extinction

Considerable attention was drawn to the phenomenon of mass extinction in the 1980s and 1990s. This attention arose from the bold assertion made in 1980 by Nobel-prizewinning US physicist Luis W Alvarez (1911–1988) and his colleagues that the dinosaurs and 70% of all other species of the Earth were wiped out 65 million years ago by the impact of an asteroid or comet measuring approximately 10 km/6 mi or more in diameter. Alvarez and his co-workers made their proposal on the basis of an unusual enrichment in the rare element iridium, an element chemically similar to platinum and gold, in a layer of clay marking the boundary between the Cretaceous and Tertiary geological Periods, the K–T boundary. Since their original report, the iridium anomaly, as it has come to be known, has been found in rocks and sediments deposited at the K–T boundary around the world. In 1991 a circular impact structure of K–T age was found buried beneath 1 km/0.62 mi of carbonate rock in Mexico's Yucatan peninsular. The structure, the Chicxulub crater, is the best candidate for the K–T impact site envisioned by Alvarez.

Despite its obvious attraction to scientists and nonscientists alike, the Alvarez impact hypothesis is not without problems, and there is another catastrophe that might well explain the K–T mass extinction. Massive volcanic eruptions in what is now the Deccan plateau region of western India coincided with the K–T extinction. The environmental effects of large-scale volcanism are every bit as menacing to life as those of an asteroid collision.

New theoretical studies suggest that mass extinctions are to be expected as a consequence of the dependence of living organisms on one another. Mathematical models constructed on this basis successfully mimic the pattern of extinctions over geological time, and suggest that mass extinctions are caused when vulnerability to extinction by normal evolutionary change is accompanied by unusual environmental stress imposed by catastrophes. Most scientists agree that a large asteroid-like body struck the Earth near the present-day Yucatan peninsula during K–T boundary. There is less agreement as to whether this event by itself killed the dinosaurs.

It is estimated that, on average, asteroids or comets measuring 5–10 km/3–6 mi in diameter strike the Earth every 50–100 million years. Debate surrounding the Alvarez hypothesis has heightened awareness of the influence of asteroid or comet impact on Earth's geological history.

Life in extreme conditions

Space exploration, biology, and geology joined forces several times in the 20th century. The return of samples from the Moon beginning in 1968 resulted in an infusion of funding for the development of equipment capable of detecting living organisms that might exist at extreme conditions. With the suggestion in 1997 (and later largely discredited) that meteorites that probably come from Mars may show evidence for microbial life, the scientific community is again tooling up to intensify the search for life at extreme conditions.

What is meant by life at extreme conditions? In 1977 scientists from the project FAMOUS (French–American Mid-Ocean Undersea Study) set out to observe firsthand the process of seafloor spreading. Upon descending in their deep-sea submersible vehicle, *ALVIN*, to depths of 2–3 km / 1–2 mi, the scientists were startled to find a host of new and strange life forms. These creatures were concentrated near undersea hot springs heated by magmatism along seafloor spreading centres. Perhaps the strangest of these creatures were the large red and white tube worms. Subsequent studies showed that these life forms owe their survival to symbiosis with bacteria that thrive on the hydrogen sulphide gas (familiar to many of us by its rotten egg odour) emanating from the volcanic hot springs. Here, the energy for life comes primarily from volcanic gases rather than from sunlight. Discovery of these organisms showed that life could exist at higher temperature, higher pressures, and with less light than previously thought, at extreme conditions.

If life existed on Mars, and this is far from proven, then could it still exist today beneath the surface of the planet? Answers to these questions begin at home. It is not yet known with certainty how deep into Earth's regolith life exists, nor is the maximum temperature at which microbial life can exist on Earth. Terrestrial geologists, planetary geologists, microbiologists, and biochemists are actively pursuing answers to these questions.

Tube worms Riftia pachyptila *at a Pacific hydrothermal vent. Tube worms can grow up to 2m/6 ft in length and were unknown to science until the later 1970s.* OAR/National Undersea Research Program (NURP)

Overview

Plate tectonics

Many of us at one time or another have glanced at a globe or a map of the world and noticed that the eastern edge of South America fits neatly with the western edge of Africa. So too have many serious thinkers before us. And many of us have been exposed to the idea that there once was a great continent, and that the parallelism of the continental coastlines is no accident. In hindsight this seems obvious. But it is useful to explore this form of logic further in order that we might better understand the boldness required to suggest that the continents of Africa and South America, for example, were indeed once attached.

Although several scientists of the 19th and early 20th centuries had suggested that the continents were once joined and then drifted apart, German meteorologist Alfred Wegener (1880–1930) is credited with having put forward a coherent theory for continental drift backed by the strength of his conviction. He suggested that approximately 200 million years ago a single great continent, which he called 'Pangaea', began to break apart. Wegener was influenced not only by the jigsaw fit of coastlines, but also by palaeontological evidence suggesting that there must have been a bridge of land between Brazil and Africa, and between Australia, Africa, and India. In addition he cited abundant similarities in the geological histories recorded by rocks on either side of the Atlantic. He put forth his ideas on 'Die Verschiebung der Kontinente', or 'continental displacement', first in a lecture in 1912 and then in a series of publications from 1913 to 1924.

Perhaps the greatest of all insults to a scientist is to be ignored. If so, Wegener's hypothesis landed him many a compliment. His palaeontological and geological evidence was scrutinized and seemingly refuted. A few prominent geologists could see merit in Wegener's 'continental drift', as it came to be known in English. But Wegener could not come up with a realistic driving mechanism for continental drift, and the theory was rejected by many on that flaw alone.

In a remarkably prescient work in 1929, English geologist Arthur Holmes (1890–1965) described a mechanism for continental drift that bears many similarities to present-day plate tectonics. Holmes had studied radioactivity in rocks for years, and it was well known that radioactive decay causes rocks to heat up. He used his knowledge of the amount of heat in the Earth to surmise that at great depth rocks must 'flow' much like the way hot water rises, cools, and then sinks in a boiling pot, a process known as convection. Holmes argued that the Earth was layered. The upper brittle portion composed of rocks like granite was underlain by a 'fluid' layer that, given sufficient time, would flow. A useful analogy can be made between flow of Holmes's 'fluid' layer over millions of years and the way in which old window panes flow, and thus thicken at their base, over hundreds of years.

This so-called fluid layer was composed, it was asserted, of peridotite, a rock denser than granite.

Since continents were known to be rich in radioactive elements like uranium, Holmes concluded that temperature was highest beneath continents. The hot rock beneath was apt to rise during convection and thus force the continents apart. Conversely, at the margins of continents the underlying rock would tend to sink, explaining the presence of deep oceanic trenches around the rim of the Pacific, for example. With his model Holmes was able to offer cogent explanations for many geological phenomena, including mountain building and volcanism in Japan, the Aleutian Islands, and around the rest of the Pacific margin.

Lack of new information slowed the pace of ideas concerning continental drift until after World War II. As a result of the war, sonar became available to better map the topography of the ocean bottom. Devices called magnetometers, perfected for use as submarine chasers, could be used to detect more accurately variations in Earth's magnetic field. By the mid-1950s these technological advances afforded new data that begged explanation.

Mapping of the ocean floor showed that there was a semicontinuous mountain range, or mid-ocean ridge, 50,000 km/30,000 mi long beneath the seas. The average elevation of this mountain chain above the seafloor is, it was discovered, nearly as high as the tallest mountain in North America. The ridges spew large amounts of heat and are rocked by earthquakes.

Deep canyons at the bottom of the sea were discovered too. These deep-sea 'trenches' were found concentrated along the margins of the Pacific Ocean basin seaward of the many volcanic centres that comprise the so-called ring of fire, including the Philippine and Aleutian Islands and Japan. The trenches are also seismically active.

Surveys of rock magnetism showed that either Earth's magnetic field or the rocks that record it were moving across the globe with time. Meanwhile, the evidence for periodic reversals in Earth's magnetic field was mounting; by accurately dating magnetized rocks using the amount of radioactive decay of potassium (K) to form argon (Ar), it was shown that rocks of similar age from widely separated regions showed the same magnetic polarity.

Seafloor spreading

Princeton University geologist Harry H Hess (1906–1969) put forth a powerful explanation for the new post-war observations in a widely circulated but unpublished manuscript in 1959. During World War II, Hess captained a US Navy ship in the Pacific. Between battles, he conducted echo soundings of the seafloor and his attention was drawn to explaining the structure of the ocean basins. His conclusion, elaborated in a landmark paper in 1962 entitled 'History of Ocean Basins', was that the ocean crust, made of basalt, a dark-coloured volcanic rock, was like a giant conveyor belt. It was produced by volcanism at the mid-ocean ridges, pushed or pulled away from the ridge axis, and eventually destroyed by plunging down into the deep-sea trenches. Hess's hypothesis, influenced by the earlier concepts of Arthur Holmes and referred to as 'seafloor spreading', was embraced by the scientific community.

Hess's theory of seafloor spreading, caught on in a way that continental drift never did. In 1963 Canadian L W Morley (1920–) and, independently, Fred Vine (1939–1988) and Drummond Matthews (1931–) of Cambridge University showed that reversal of Earth's magnetic field in combination with seafloor spreading could account for the 'stripes' of magnetic polarity recorded in basaltic rocks of the ocean bottom. These stripes are symmetrically disposed about the mid-ocean ridges, and could be explained by cooling of the newly made volcanic basalt as it moved away from the axis of the ridge. As the rock cools, it takes on the magnetic field of the Earth at that time, and so forth.

In 1965, Canadian geologist J Tuzo Wilson (1908–1993) elaborated on Hess's seafloor spreading idea. He suggested that the mid-ocean ridges, deep-sea trenches, and the faults that connect them combine to divide the Earth's outer layer into rigid independent plates. The connecting faults were shown by Wilson to have a unique geometry, and were called transform faults.

The formal theory of plate tectonics was enunciated in the late 1960s by Jason Morgan of Princeton University, Dan McKenzie of Cambridge, and Xavier Le Pichon (1937–), then at the Lamont Observatory in New York. All three had training in physics that benefited the development of the theory. Morgan used mathematical arguments to show that the rigidity of the plates requires that they extend to depths beyond the lower limit of the oceanic crust. The term 'tectosphere' was used to describe the crust and upper mantle that comprise the rigid plates. The tectosphere is now referred to as the lithosphere. Mid-ocean ridges, or spreading centres, were identified as sites of plate construction. Trenches, or subduction zones, marked the places where the plates sink back into the deep Earth. Le Pichon, McKenzie, and their colleagues demonstrated that motions of the plates made sense, geometrically, when inferred from the relative positions of the spreading centres, subduction zones, and connecting transform faults.

Geophysicists like McKenzie and Don Anderson (1933–) of the California Institute of Technology used mathematics to investigate possible driving forces for the plates. Again, the ideas of Arthur Holmes proved prescient. Convection in the mantle, whereby hot rock creeps slowly upward, spreads latterly, and then cools and descends, seemed the best candidate. Upward convection was postulated to be occurring beneath the mid-ocean ridge system while downward convection explained the sinking of the plates at deep-sea trenches.

We now know that the motions of the plates of the lithosphere are aided by the presence of a mobile layer of the mantle beneath them, known as the asthenosphere. The asthenosphere, defined by the slowing of seismic waves as they pass through it, is remarkably similar to Harry Hess's 'fluid' layer.

The word tectonic connotes assembly, and plate tectonics derives its name in part because the distribution of mountain ranges is explained by the motions of the plates of lithosphere. For example, as the Indian plate moved northward the continental crust riding passively on it eventually encountered the continental crust of the Eurasian plate. Both continental masses are too buoyant to be pulled down with the remainder of the subducting lithosphere and so they are squeezed together. Thus the Himalayan mountain range was born.

Beyond plate tectonics

Beginning in about 1980, many geologists, and particularly those working in western Canada, began to find evidence that some mountain ranges are composed of numerous small pieces of crust that originate from far-off places. A new terminology ensued. Blocks of crust that are bounded by faults on all sides were labelled suspect terranes by some scientists because their place of origin was often in doubt, or suspicious. If and when a fault-bound block is shown to be far travelled, its status is upgraded from suspect terrane to simply terrane. The terrane concept predicts that Borneo, Sumatra, and Java may be future terranes to be imbedded into the margin of a larger continent following millions of years of plate motions.

Although it explains the positions of the mountain ranges, the ways in which continents deform to form mountains are not explained by plate tectonics. Fundamental observations like the heights of mountains on continents remain unexplained. Today, researchers are using advancements like the global positioning system (GPS) to measure the deformation of continents with great precision. The results of such studies may one day be useful for identifying areas on the continents that are susceptible to earthquakes.

Hazard assessment

In the last century, earth scientists have been increasingly called upon to assess natural hazard risks and to help local authorities plan and implement strategies for dealing with those risks. In 1902, Mt Pelée on the Island of Martinique, West Indies, erupted, killing all but two of the 28,000 inhabitants of the city of St Pierre. In April 1906, a powerful earthquake and ensuing fire destroyed the city of San Francisco and killed at least 3,000 people. These sorts of disasters emphasized the need to understand the underlying causes of such events in order to predict them and warn inhabitants.

Although it is still nearly impossible to predict earthquakes, seismologists now have a very strong understanding of where earthquakes are likely to occur. They have also made tremendous strides in understanding why earthquakes happen, from the global scale (plate tectonics) down to the microscopic. Understanding how rocks break, and how the resulting seismic waves are transmitted through the Earth and over its surface, have allowed them to draw up earthquake hazard maps, and to liase with engineers to design structures that are more earthquake resistant. Although it is impossible to prevent naturally occurring earthquakes, scientists have, in fact, used their knowledge of fault mechanics to prevent earthquakes induced by human activities such as pumping water and building dams.

During the 20th century, predicting volcanic eruptions has become much more of an exact science. From detailed field and laboratory studies of volcanoes, both active and extinct, over the past 100 years, geologists now have a very clear understanding of how volcanoes erupt and where they are likely to erupt. Volcanologists now use a variety of tools in predicting eruptions. Detailed field studies show the frequency, duration, and intensity of volcanic events in the past; seismometers record tremors possibly related to magma movement; satellites can monitor heat, ash, and some gas output as well as the physical changes of an inflating volcano.

However, in spite of the ability of geologists to accurately predict volcanic eruptions at volcanoes that are monitored, most volcanoes are not monitored. This is for a variety of reasons, money and politics being the most important, and stresses the fact that natural hazard mitigation has as much to do with politics as with science.

Of course, other than pure curiosity about what is in store for the near future, meteorology too is driven by risk mitigation. Meteorological hazards such as storms and floods are more dangerous – being more frequent and affecting more people – than either earthquakes or volcanoes. Advances in both understanding atmospheric dynamics and the interactions between the atmosphere and landscape, as well as advances in remote sensing and observing these dynamics and interactions have saved thousands of lives over the past few decades.

Coupled with natural hazard mitigation is environmental geology, a field that took off in the last thirty years or so of the 20th century. Geochemists, for instance, have turned their studies towards understanding the chemistry of radon and of pollutants such as lead and mercury.

It is these sorts of applications that justify basic research and drive funding. In general, the public is more willing to part with its money if it knows it will be applied to something directly concerned with its welfare, rather than to an obscure academic problem. The increasing focus on hazard assessment has served the dual purpose of saving lives and increasing funds for basic research.

Remote sensing

Remote sensing has allowed many natural hazards to be monitored in a way never before available to us. Storm systems, for example, can be tracked hour by hour. Yet the significance of remote sensing extends far beyond hazard assessment and even meteorology. The advent of remote sensing technologies, primarily in the second half of the 20th century, has afforded earth scientists completely new views of the Earth. First balloons, then aeroplanes, and finally satellites, broadened our scope, allowing scientists to observe landforms on a scale that had only previously been possible with meticulous mappers and very skilled artists. Innovations in remote sensing, which were initially developed for military applications, made certain kinds of mapping cheaper, faster, and more accurate. They have allowed geologists to visit, albeit remotely, terrains that had been physically (or politically) out of reach. Furthermore, remote sensing technologies have allowed scientists to observe Earth *processes* in a way they had never before been able. With a number of images taken over a series of days, months, or years, for instance, or with near-continuous monitoring, earth scientists can easily document changes in the Earth's landscape or atmosphere and more easily observe the processes responsible for those changes.

Prior to the 1950s, we were only able to view the visual part of the electromagnetic spectrum, that is, visible light, or the wavelengths of electromagnetic energy that are visible to the human eye. Sensors are now able to detect the entire range of electromagnetic radiation, from gamma rays through visible light to microwave radiation, that is reflected or emitted from the Earth.

Every material absorbs and emits radiation differently. For instance, plants appear green because their chlorophyll absorbs red and blue wavelengths, and reflects green; water appears blue because it reflects blue wavelengths, and absorbs the others. Two materials that appear the same in their visual spectra, however, may have very different infrared or ultraviolet properties. Infrared sensors, for instance, 'see' in the infrared spectrum, and can therefore discriminate materials that absorb and reflect infrared radiation differently (even if they look the same in visible light, that is, they are the same colour).

This sort of technology has been used to differentiate soils, rocks, and vegetation for such applications as geological mapping, mineral prospecting, and land use history and planning. Sensors that detect thermal infrared radiation are used to map sea surface temperature, to prospect for geothermal energy sources and near surface groundwater, and to monitor volcanic activity. Remote sensing has also revealed structures such as faults that have been invisible because no geologists had conducted field work in the area or simply because we could never stand back far enough to see them.

Although the term 'remote sensing' is generally used to refer to imaging the electromagnetic spectrum from far above the Earth's surface, it can really refer to any sort of observation made from a distance. In the early part of the 20th century, earth scientists were already conducting remote sensing surveys of the ocean floor. Using shipboard gravity meters, magnetometers, and sonar ranging instruments, they remotely mapped the seafloor. Without these remote sensing technologies, our view of the Earth's surface, especially the two-thirds that lies underwater, would be very vague. It was remote sensing that revealed the details of the seafloor and enabled the theory of continental drift to develop into plate tectonics.

In the latter part of the 20th century, satellite-borne remote sensing devices have given us an even more precise view of the subaerial topography of the planet. Laser ranging instruments have enabled us to make topographic maps of the Earth accurate to within a few centimetres. Using these techniques, Mount Everest, for instance, has been measured at 8,850 m/2,900 ft above sea level. Lasers also measure the height of the oceans, giving us information on ocean temperature, and even ocean bathymetry (ocean height is related to gravity, which is in turn related to mass. Because rock is more massive than water, gravity is high over submarine mountains and low over trenches). Satellite gravity data have given us maps of regions even more remote than the nearside of the Moon: the seafloor.

The importance of remote sensing to plate tectonic theory cannot be overstated. Not only have we been able to gather incredible amounts of information on the physical structure of the Earth's surface, with these technologies earth scientists are actually able to measure the vertical and horizontal movement of points on the continents from one year to the next. We have now observed that every year, Mount Everest grows 4 mm/0.16 in taller and travels 5 mm/0.20 in towards the northeast.

Remote sensing has also, of course, changed the face of meteorology. Using geostationary satellites that 'hover' 36,000 km/22,370 mi high, imaging one-third of the Earth every half-hour, meteorologists are able to observe the atmosphere on a spatial and temporal scale they had never before been able.

Like planetary geology, it was advances in engineering that allowed geologists to begin observing the Earth from far away. Along with spacecraft and sensor engineering, computers have given us the capability to use the data efficiently and effectively.

Planetary geology

At about the time of the start of the plate tectonic revolution, another innovation in earth science was brewing: planetary geology. Planetary geology had its roots in the late 19th century. Grove Karl Gilbert (1843–1918), the USA's finest field geologist, surveyed the Moon through a telescope for several months, and then wrote 'The Moon's Face', a paper discussing the origin of its craters. At the time, almost everyone assumed the craters were volcanic in origin, but Gilbert, having experimented with creating impact craters by dropping metal spheres into mud, argued that they formed when objects smashed into the Moon. Several years later, Gilbert investigated Coon Butte crater (now known as Meteor Crater or the Barringer Crater) in Arizona. He was sure that it was an impact crater similar to those on the moon, but could never find enough evidence to prove it to himself or anyone else.

The recent geological (as opposed to astronomical) study of the Solar System's other planetary bodies was spearheaded by US Geological Survey (USGS) geologist Gene Shoemaker (1928–1997). In the 1950s Shoemaker began investigating the Barringer Crater, and eventually proved that it was in fact formed by a meteorite impact rather than a volcanic explosion.

His interest in impact piqued, Shoemaker went on to look at the craters on the Moon. He and his colleagues applied the basic tenets of terrestrial field geology to 'astrogeology' as the field was known then. Using photographs of the Moon (first from telescopes, later from lunar probes) he used the principles of superposition and cross-cutting relationships to map small areas of the Moon, making a first determination of its geologic history. Years before anyone set foot on the Moon, geologists had determined a significant amount about its geology. Many geologists feel that without the determination of Shoemaker to make the Moon programme a scientific as well as an engineering and political venture, there would be no discipline of planetary geology as we know it today.

Understanding the origin of lunar craters was fundamental to understanding the history and evolution of the planets, including Earth. Impact craters (along with the *Apollo* samples, which were dated radiometrically) showed that the Moon had been continuously bombarded during its first 700 million years or so. Not only did this give greater credence to the idea that planets formed though the combination of smaller planetesimals and the accretion of rocks in space, it also provided a means for dating surfaces of other planets. The basic idea is that the longer a surface has been exposed, the more impact craters it will have accumulated. Ancient terrains will hold the most, and the largest craters, while landscapes that have been resurfaced more recently (either by erosion or deposition) will have fewer craters. Impact craters thus provided a means of determining the chronology of events and the geological history of the surfaces of other planets.

An additional lesson that came out of the recognition of lunar craters as impact had to do with the ongoing push and pull between uniformitarianism and catastrophism. The foundation of modern geology (that is, geology since the late 18th century) lies in the 'principle of uniformitarianism': the idea that the processes, and rates of processes, that shaped the Earth in the past are the same as those in the present. This was contrasted with 'catastrophism', the idea that Earth's surface is the result of a few major catastrophic events, such as the Biblical Flood. Uniformitarianism was first applied to geology in a formal way by James Hutton in the late 1700s, and was championed by Charles Lyell in the 1830s. Since its general acceptance, most geologists have been loath to invoke catastrophic mechanisms to explain features in the geological record. Accepting that there were impacts on the Moon, and that Meteor Crater was an impact crater, meant accepting that 'catastrophic' events can, in fact, shape a planet. Without this, it is unlikely that geologists would have accepted Alvarez's hypothesis that a giant impact was responsible for the extinctions at the end of the Cretaceous period.

The data gathered from photographs, field trips, and samples of the Moon also served to constrain its history and origin. The currently accepted theory (although it still leaves some observations about the Moon unclear) is that the Moon formed soon after the Earth, from the debris of a collision between a large planetesimal and the Earth.

Beyond the Moon

The missions to the moon were just the beginning of geological exploration of the planets. The Soviet *Mars* missions and US *Mariner* and *Viking* missions in the 1970s revealed Mars to be a cold desert, but one that had obviously sustained flowing water in the past. Larger than the Moon, Mars held its accretionary heat for longer, was volcanically active longer, and generally had a more complicated geological history. Although it is marked with tectonic features such as volcanoes, faults, and a giant rift valley, Mars does not appear to have undergone plate tectonics. The incredibly high resolution photography of the *Mars Global Surveyor* in the last few years of the century has added additional, more detailed information about the history of the planet. The US *Pathfinder* mission, with its rover 'Sojourner', was a further step towards first-hand sampling (as opposed to sampling via meteorites) of the planet.

The 1970s Soviet *Venera* mission to Venus showed that Earth's 'twin' was not a tropical jungle, as many had hoped, but a searing hot landscape with an atmosphere 30 km/19 mi thick and 90 times heavier than Earth's. The US *Pioneer Venus* and *Magellan* missions used radar to probe the surface beneath Venus's thick atmosphere. Unlike Mercury, Mars, and the Moon, Venus is not covered in impact craters. It certainly contains many more pristine impacts than the Earth, but there are still fewer than a thousand. In addition, there are no recognized craters less than about 1,500 m/1,640 ft across.

Although the lack of small craters has a simple explanation – small impactors burned up in the thick atmosphere – the paucity of larger craters has a more interesting one. Like the Earth's, Venus's surface must have been very active in order to obliterate so many craters. Volcanoes, ridge belts, fracture and rift zones also point to a tectonically active past. The nature of the tectonics, however, whether plate tectonics, micro-plate tectonics, or vertical tectonics, is still not understood.

Although natural satellites are not considered to be planets in the astronomical sense, if they are spherical they are, geologically speaking, planets. Most asteroids and nonspherical satellites are essentially big rocks, and therefore also of interest to the geologist. The first really useful images of the satellites of the outer Gas Giants came from the *Voyager* spacecrafts, which were launched in 1977. What came back to Earth were images of a completely unexpected variety of landscapes: the active sulphur volcanoes of Io, the icy surface of Europa, the huge craters of Tethys and Mimas, the global grooves of Enceladus, and the 'cantaloupe terrain' of Triton. The *Galileo* mission of the late 1990s served to sharpen our view of the satellites. Then, in the late 1990s, the NEAR (Near Earth Asteroid Rendezvous) mission brought us more detailed information about asteroids, the bodies that most meteorites derive from.

Many lessons have come out of planetary exploration. We've observed that on the one hand, there are many geological processes that are common to the history of a number, if not all, planetary bodies (such as volcanism and cratering). But what exploration has also showed us is that although there are common processes, the combination of processes, the magnitude and rate of those processes, the composition, and the history of each planetary body appear to be unique. There is an incredible variety of terrains and landforms in the Solar System, a variety that no one imagined prior to the Space Age.

The Earth is, in fact, as far as we know, unique. It is the only one, so far, that has been shown to have sustained plate tectonics. It is the only planet that we know contains water as a solid, a liquid, and a gas. And it is the only planet we know of that contains extant life. Planetary geology has given us a much better understanding of the Earth as a planet and a geological system.

2 Chronology

1899–1900
Serbian-born US inventor Nikola Tesla discovers stationary terrestrial waves, demonstrating that the Earth can act as a conductor of certain electromagnetic frequencies.

1900
French meteorologist Léon-Philippe Teisserenc de Bort discovers that the Earth's atmosphere consists of two main layers: the troposphere, where the temperature continually changes and is responsible for the weather, and the stratosphere, where the temperature is invariant.

German meteorologist Vladimir Peter Köppen develops a mathematical system for classifying climatic types, based on temperature and rainfall. It serves as the basis for subsequent classification systems.

Swedish chemist Per Teodor Cleve publishes *The Seasonal Distribution of Atlantic Plankton Organisms*, which serves as a basic text on oceanography for many years.

1902
Italian physicist Guglielmo Marconi discovers that radio waves are transmitted further at night than during the day because they are affected by changes in the atmosphere (actually by a layer of ionized gas in the ionosphere).

1903
The US president Theodore Roosevelt establishes the first US national wildlife refuge on Pelican Island, off the east coast of Florida. By 1929, 87 federal refuges will be established.

1906
Irish geologist Richard Oldham proves that the Earth has a molten core, by studying seismic waves.

1907

German scientists Richard Anschütz and Max Schuler perfect the gyrocompass, which always points to true north.

1908

13 May
44 state and territorial governors attend an environmental conservation conference, convened by US president Theodore Roosevelt, at the White House in Washington, DC.

1909

Croatian physicist Andrija Mohorovičić discovers the Mohorovičić discontinuity in the Earth's crust. Located about 30 km/18 mi below the surface, it forms the boundary between the crust and the mantle.

US geologist and secretary of the Smithsonian Institution in Washington, DC, Charles D Walcott discovers fossils of soft parts of organisms in the Cambrian Burgess Shale of the Canadian Rockies. The discovery provides unprecedented evidence pertaining to the rapid evolution of life that started in the Cambrian period.

August
Seattle, Washington, hosts the first National Conservation Congress, representing 37 states.

1910

Swedish archaeologist Gerhard de Geer publishes *A Geochronology of the Last 12,000 Years*, setting out his influential system for dating rock strata.

US physicist Percy Bridgman invents a device, called the 'collar', that allows him to squeeze all kinds of materials to pressures comparable to the base of Earth's crust, giving rise to the new fields of high-pressure physics and mineral physics.

1911

Austrian physicist Victor Francis Hess discovers cosmic radiation using crewed balloons.

1912

German meteorologist Alfred Wegener suggests the idea of continental drift and proposes the existence of a supercontinent (Pangaea) in the distant past.

1913

English geologist Arthur Holmes uses radioactivity to date rocks, establishing that the Earth is 4.6 billion years old.

French physicist Charles Fabry discovers the ozone layer in the upper atmosphere.

1914

German-born US geologist Beno Gutenberg discovers the discontinuity that marks the boundary between the Earth's lower mantle and outer core, about 2,800 km/1,750 mi below the surface.

1915

Finnish chemist and geologist Pentti Eelis Eskola develops the concept of metamorphic facies.

1919

Norwegian meteorologists Vilhelm and Jakob Bjerknes introduce the term 'front' in meteorology (after the military front), which describes the transition zone between two masses of air differing in density and temperature.

The US Navy develops the Hayes sonic depth finder. It consists of a device that generates sound waves and receives their echo from the ocean floor. A timing device indicates the depth of the water.

1920

English physicist Frederick Soddy suggests that isotopes can be used to determine geological age.

Yugoslavian meteorologist and mathematician Milutin Milankovitch shows that the amount of energy, or heat, received by Earth from the Sun varies with long-term changes in Earth's orbit. Decades later, scientists will correlate fluctuations in global temperature to his 'Milankovitch cycles'.

1921

Norwegian meteorologist Vilhelm Bjerknes publishes *On the Dynamics of the Circular Vortex with Applications to the Atmosphere and to Atmospheric Vortex and Wave Motion*, in which he summarizes his work on the movement of air masses and weather forecasting.

The crew of the brigantine Carnegie, *a research vessel, stands in front of the binnacle, a glass dome protecting the ship's compass (c. 1920). The* Carnegie *conducted global magnetic, electric, and gravity surveys.* Corbis/The Mariner's Museum

1922

British meteorologist Lewis Fry Richardson publishes *Weather Prediction by Numerical Process*, in which he applies the first mathematical techniques to weather forecasting.

Geophysics is used for prospecting for the first time. The torsion balance, a device for measuring gravitational field strength, is used to locate salt domes, and thus oil, in the Gulf of Mexico.

1923

German-born British physicist Frederick Lindemann investigates the size of meteors and the temperature of the upper atmosphere.

The first seismic prospecting takes place in the USA when geologists use seismometers to discover an oil field.

1924

US Congress passes the Oil Pollution Act, prohibiting oil producers from polluting the environment.

1925

The US Navy develops a pulse modulation technique to measure the distance above the Earth of the ionizing layer in the atmosphere.

1926

The Scott Polar Research Institute is opened in Cambridge, England, to conduct Antarctic research.

1928

Canadian geochemist Norman Levi Bowen publishes 'The Evolution of the Igneous Rocks' in which he suggests that Earth's crust is the product of melting of parts of the mantle, a process known as differentiation. His work firmly establishes the potential of the physical chemical approach to geology.

1929

By studying the magnetism of rocks, the Japanese geologist Motonori Matuyama shows that the Earth's magnetic field periodically reverses direction.

English geologist Arthur Holmes describes a mechanism for continental drift that bears many similarities to present-day plate tectonics.

Fire destroys the research sailing vessel *Carnegie* in Samoa following a gasoline explosion after the ship had logged 548,290 km/342,681 mi of ocean during magnetic, electric, and gravity surveys of the globe.

French engineer and chemist Georges Claude demonstrates that the temperature difference between the upper and lower depths of the ocean can be used to generate electricity.

Norwegian chemist Victor Moritz Goldschmidt produces the first table of ionic radii which is useful for predicting crystal structures.

c. 1930

Explosion seismology – the study of seismic waves caused by explosions – is used by the oil industry to explore for oil in the USA.

1930

The Woods Hole Oceanographic Institution is established in Massachusetts.

11 June
The first bathysphere, a spherical steel craft for undersea exploration, built by US zoologist William Beebe and US engineer Otis Barton, descends to 435 m/1,428 ft.

December
Hundreds of people fall ill and 60 die during a four-day fog in the industrialized Meuse Valley in Belgium. It is the first recorded air pollution disaster.

1934

18 August
US explorers and biologists William Beebe and Otis Barton descend in a bathysphere to a record 923 m/3,028 ft in the Atlantic off Bermuda.

US zoologist William Beebe and US engineer Otis Barton descend in a bathysphere to a record 923 m/3,028 ft in the Atlantic off Bermuda in 1934. Corbis/Ralph White

1936

Russian geochemist D S Korzhinsky publishes an extension of thermodynamics that accounts for reactions between rocks and fluids flowing through them.

The Danish seismologist Inge Lehmann postulates the existence of a solid inner core of the Earth from the study of seismic waves.

The radio meteorograph (radiosonde) is developed by the US Weather Service; it transmits information on temperature, humidity, and barometric pressure from uncrewed balloons. A network of stations is also inaugurated.

1937

British physicist Lawrence Bragg's book *Atomic Structure of Minerals* is published.

Finnish chemist and geologist Victor Moritz Goldschmidt tabulates the absolute abundances of chemical elements in Earth from solar and meteorite chemical data.

31 October
Sphinx Rock meteorological station in Bernese Oberland, Switzerland, is opened.

1938

Seismologists Beno Gutenberg and Charles Richter report the deepest earthquake shock on record; it occurred in 1934 at a depth of 720 km/450 mi beneath the floor of the Flores Sea, southern Indonesia.

1939

The US geophysicist Walter Maurice Elsasser formulates the 'dynamo model' of the Earth, which proposes that eddy currents in the Earth's molten iron core cause its magnetism.

c. 1940

Information from uncrewed weather balloons indicates that columns of warm air rise more than 1.6 km/1 mi above the Earth and winds form layers in the lower atmosphere, often blowing in different directions.

1945

Single-stage sounding rockets, reaching speeds of 4,800–8,000 kph/3,000–5,000 mph, and a maximum altitude of 160 km/100 mi, are launched carrying instrumentation to gather information about the upper atmosphere.

1946

The first cloud-seeding experiments are conducted in the USA in an attempt to produce rain.

1947

The US meteorologist Irving Langmuir carries out the first hurricane-seeding experiment; 91 kg/200 lb of dry ice is distributed in a storm.

1948

US chemist Harold C Urey, in a seminal paper, foretells the use of stable isotopes as 'geothermometers' and chemical tracers in the earth sciences.

US chemists Lyman T Aldrich and Alfred Nier find argon from decay of potassium in four geologically old minerals, confirming predictions by German physicist Carl Friedrich von Weizsacker made in 1937. The basis for potassium-argon dating is established.

1949

English geophysicist Edward C Bullard and colleagues design a probe for measuring Earth's heat flow through the ocean floor (published in 1952).

Japanese metamorphic petrologist Akiho Miyashiro estimates the pressure and temperatures for the aluminosilicate polymorphs (sillimanite, kyanite, and andalusite).

US seismologist Hugo Benioff identifies planes of earthquake foci, extending from deep ocean trenches to beneath adjacent continents at approximately 45°, as faults. These faults, later called Benioff zones or Benioff-Wadati zones, will later be identified as the tops of plates of lithosphere.

1950

The Hungarian-US mathematician John von Neumann makes the first 24-hour weather forecast using a computer.

1952

The Norwegian-US meteorologist Jacob Bjerknes discovers that centres of low pressure, or cyclones, develop at the fronts that separate different air masses. It leads to improved weather forecasting. He is also the first to use photographs, taken from high-altitude rockets, as a tool in weather analysis and forecasting.

US geophysicist Francis Birch predicts fundamental changes in the mineralogy of Earth's mantle with depth.

US meteorologists establish the first weather station in the Arctic, at Ice Island T-3.

1953

US chemist Loring Coes invents an apparatus for obtaining high temperatures at high pressures in the laboratory. He uses this apparatus to grow high-pressure minerals, including a new form of dense silica that bears his name, coesite.

US geophysicist William Ewing announces that there is a crack, or rift, running along the middle of the Mid-Atlantic Ridge.

1 August
The US bathyscaph *Trieste* is launched. Later in the year it dives to a record 3,150 m/10,300 ft.

1954

Researchers at the US electrical company General Electric produce the first synthetic diamonds.

The existence of pre-metazoan life becomes widely accepted when diverse fossil microscopic organisms are discovered in the Gunflint rocks along the north shore of Lake Superior. The Gunflint biota prove that there had been significant evolutionary activity by at least *c.* 2 billion years before present.

11 January
George Cowling of the Meteorological Office becomes the first weather forecaster to appear on British television. Forecasts had previously been done by voice-over.

15 February
The French bathyscaph *FNRS 3* descends to a record 4,000 m/13,000 ft in the Atlantic Ocean off Senegal.

October
A US Aerobee rocket takes the first picture of a complete hurricane, at an altitude of 160 km/100 mi off the Texas Gulf coast.

1955
April
The US communications engineer John Pierce analyses various types of satellites and shows the potential of using the Earth's gravity to control the altitude and orientation of satellites in geosynchronous orbit (i.e. orbits that lie at the same distance from Earth but are inclined to the equator). It leads to the launch of the communications satellite *Telstar*.

1956
Geochemist C C Patterson uses the isotopic composition of lead to determine that three stoney meteorites and two iron meteorites have a common age of 4.55 billion years. Patterson uses lead isotopes to establish that Earth is the same age.

The Clean Air Act is passed by the British Parliament. It prohibits the burning of untreated coal in London, England, and successfully reduces the emission of sulphur-oxide pollution.

US geologists Bruce Heezen and William Ewing discover a global network of oceanic ridges and rifts 60,000 km/37,000 mi long that divide the Earth's surface into 'plates'.

1957
Australian meteorologists create artificial rain in New South Wales, Australia, increasing rainfall by 25%. In Queensland it saves crops.

1 July
International Geophysical Year begins. A cooperative international research programme, it involves scientists from 70 nations in Antarctic exploration, oceanographic and meteorological research, geomagnetism, seismology, and the launching of satellites into space.

1958
The Equatorial Undercurrent is discovered in the equatorial Pacific Ocean. It has a width of 320–480 km/200–300 mi, a height of 200–300 m/650–1,000 ft, and flows 50–150 m/165–500 ft below the surface.

17 March

The USA launches *Vanguard 1*. The second US satellite, it tests solar cells and consists of a 1.47 kg/3.25 lb sphere equipped with two radio transmitters. It proves that the Earth is slightly pear-shaped.

May

Using data from the *Explorer* rockets, US physicist James Van Allen discovers a belt of radiation around the Earth. Now known as the Van Allen belts (additional belts were discovered later), they consist of charged particles from the Sun trapped by the Earth's magnetic field.

15 May

The USSR places *Sputnik 3* in orbit. It contains the first multipurpose space laboratory and transmits data about cosmic rays, the composition of the Earth's atmosphere, and ion concentrations.

27 August

The USA conducts the first Argus experiment, the explosion of three high-altitude nuclear bombs to study the effect of the Earth's magnetic field on the charged particles released by the explosions.

1959

Russian D S Korzhinskii's new rules for the thermodynamics of open systems are published in English in *Physicochemical Basis of the Analysis of the Paragenesis of Minerals* (translated from the Russian version of 1957).

US geologists Marion Hubbert and William Rubey demonstrate that the overthrusting of large horizontal planes of rock that produces folded mountains is due to the reduction of friction caused by fluids in the rocks.

February

The US Navy launches *Vanguard 2*, the first weather satellite.

28 February

The US Air Force launches *Discoverer 1* into a low polar orbit where it photographs the entire surface of the Earth every 24 hours. The exposed film is returned to Earth in its ejectable capsule.

April

The US Naval Research Laboratory reports a 300% increase in atmospheric radioactivity in the wake of Soviet resumption of nuclear testing in 1958, following a three-year self-imposed moratorium.

1 December
An Antarctic Treaty is signed, suspending territorial claims and aiming to prevent development in the region (valid 23 June 1961–December 1989).

c. 1960
Meteorologists begin to study storm systems using Doppler radar, which can detect the speed and direction of moving storms because of the change in frequency of the reflected radar waves.

1960
US geophysicist Harry Hess develops the theory of seafloor spreading, in which molten material wells up along the mid-oceanic ridges forcing the seafloor to spread out from the ridges. The flow is thought to be the cause of continental drift.

23 January
Swiss engineer Jacques Piccard and US Navy lieutenant Don Walsh descend to the bottom of Challenger Deep (10,916 m/35,810 ft), off the Pacific island of Guam, in the bathyscaph *Trieste*, setting a new undersea record.

24 February–10 May
Following the route of Portuguese explorer Ferdinand Magellan, the US nuclear-powered submarine *Triton* makes the first underwater circumnavigation of the globe, travelling 61,430 km/41,519 mi.

1 April
The USA launches *TIROS 1* (Television and Infra-Red Observation Satellite). A weather satellite, it is equipped with television cameras, infrared detectors, and videotape recorders. It provides a worldwide weather observation system, along with subsequently launched *Tiros* satellites.

1961
L O Nicolaysen invents the 'isochron' method of rubidium–strontium and uranium–lead dating of geological materials.

US geophysicist Francis Birch relates seismic velocities to the density and the average atomic mass of rocks in Earth's mantle.

US researchers establish Arctic Research Lab Ice Station II (Arliss II), a drifting sea ice station.

May
The Atlas computer, the world's largest (with one megabyte of memory), is installed at Harwell, England, to aid atomic research and weather forecasting.

18–20 August
The US Navy and the US Environmental Sciences Service Administration initiate Project Stormfury, an attempt to modify hurricanes through seeding, by heavily seeding Hurricane Debbie with silver iodide. Wind speeds drop markedly.

1962

US biologist Rachel Carson, in her book *Silent Spring*, draws attention to the dangers of chemical pesticides.

US geologist Harry Hess publishes *History of Ocean Basin*, in which he formally proposes seafloor spreading, the idea that the ocean crust is like a giant conveyor belt produced by volcanism at the mid-ocean ridges, pushed or pulled away from the ridge axis, and eventually destroyed by plunging down into the deep-sea trenches. Hess's hypothesis, influenced by the earlier work of Arthur Holmes, is embraced by the scientific community.

15 January
British weather reports start giving temperatures in centigrade as well as Fahrenheit.

1963

British geophysicists Fred Vine and Drummond Matthews analyse the magnetism of rocks in the Atlantic Ocean floor, which assume a magnetization aligned with the Earth's magnetic field at the time of their creation. It provides concrete evidence of seafloor spreading.

Canadian geophysicist L W Morley discovers the significance of the 'striped' pattern of magnetism in rocks of the ocean floor (published in 1964).

Geologist Ian G Gass suggests that the Troodos Massif, Cyprus, is a fragment of Mesozoic ocean floor. It is the first recognition of the significance of ophiolites.

19 June
The US satellite *TIROS 7* (Television and Infra-Red Observation Satellite) is launched. It is used by meteorologists to track, forecast, and analyse storms.

1964

The US National Science Foundation establishes a consortium of four leading institutions, called the Joint Oceanographic Institutions for Deep Earth Sampling (JOIDES), for the purpose of drilling into Earth's ocean floor. Results from the project over the next few years will eventually help confirm the theory of seafloor spreading by establishing that the oceanic crust everywhere is less then 200 million years old.

1965

Canadian geologist John Tuzo Wilson publishes *A New Class of Faults and Their Bearing on Continental Drift*, in which he formulates the theory of plate tectonics to explain continental drift and seafloor spreading.

French oceanographer Jacques Cousteau heads the Conshelf Saturation Dive Programme, which sends six divers 100 m/328 ft down in the Mediterranean for 22 days.

NASA launches *GEOS 1* (Geodynamics Experimental Ocean Satellite). Its aim is to provide a three-dimensional map of the world accurate to within 10 m/30 ft.

The Large Aperture Seismic Array is established in Montana. The signals from 525 seismometers, dispersed over an area of 30,000 sq km/11,600 sq mi, are combined to record seismic events with a high degree of sensitivity.

1966

Geochemists A E Ringwood and A Major relate the seismic discontinuity at a depth of 400 km/250 mi to the change of the mineral olivine to the structure of the mineral spinel.

The US geologists Allan Cox, Richard Doell, and Brent Dalrymple publish a chronology of magnetic polarity reversals going back 3 million years. It is useful in dating fossils.

The US National Science Foundation puts Scripps Institute of Oceanography in charge of the JOIDES project and establishes the Deep Sea Drilling Project, or DSDP.

1967

British geophysicist D McKenzie (published with R L Parker in 1967) and US geophysicist Jason Morgan (published in 1968) describe the motions of plates across Earth's surface. Morgan calls the plates 'tectosphere'. They are later referred to as plates of lithosphere.

Geophysicist Lynn R Sykes uses first-motion seismic studies to establish that mid-ocean ridges form with offsets (sections of the ridges are displaced relative to one another along faults so that the crest of the ridge has a stepped appearance) rather than being offset later. This is a major advance in the understanding of the formation of ocean basins.

US scientists Syukuvo Manabe and Richard T Wetherald warn that the increase in carbon dioxide in the atmosphere, produced by human activities, is causing a 'greenhouse effect', which will raise atmospheric temperatures and cause a rise in sea levels.

18 March
The Liberian-registered tanker *Torrey Canyon* strikes a submerged reef off the coast of Cornwall, England, and spills 860,000 barrels of crude oil into the sea. It is the biggest oil spill to date.

1968
A borehole drilled 2,162 m/7,093 ft into the Antarctic ice at Byrd Station reveals that the bottom layers of ice are 100,000 years old.

British Petroleum workers practising methods for controlling oil pollution. Oil spills became a major environmental problem in the 1960s and 1970s as a result of oil exploration along the continental shelf and several major oil spills involving supertankers. British Petroleum

British geochemist Ernest R Oxburgh and US geodynamicist Donald L Turcotte calculate the thermal consequences of subduction.

French geophysicist Xavier Le Pichon, working at the Lamont Observatory in New York, describes the motions of Earth's six largest plates using poles of rotation derived from the patterns of magnetic anomalies and fracture zones about mid-ocean ridges.

The US research ship *Glomar Challenger* starts drilling cores in the seafloor as part of the Deep Sea Drilling Project (DSDP). Capable of drilling in water up to 6,000 m/20,000 ft deep, it can return core samples from 750 m/2,500 ft below the seafloor and is equipped with a gyroscopically-controlled roll-neutralizing system that allows it to maintain its stability in diverse weather conditions.

US scientist Elso Sterrenberg Barghorn and associates report the discovery of the remains of amino acids in rocks 3 billion years old.

1969

The Joint Oceanographic Institutions Deep Earth Sampling (JOIDES) project begins. It makes boreholes in the ocean floor and confirms the theory of seafloor spreading and that the oceanic crust everywhere is less than 200 million years old.

1970

British geologist John F Dewey with J M Bird relate the positions of Earth's mountain belts to the motions of lithospheric plates.

1972

Dennis Meadows's *The Limits to Growth* is published by the Massachusetts Institute of Technology. Based on a Club of Rome report and computer simulation, it predicts environmental catastrophe if the depletion of the Earth's resources, overpopulation, and pollution are not acted upon immediately.

The United Nations Environment Programme (UNEP) is established; its aim is to advise and coordinate environmental activities within the United Nations.

23 July
The USA launches *Landsat 1*, the first of a series of satellites for surveying the Earth's resources from space.

1974

Earth scientist John Liu discovers that the lower mantle, comprising most of the Earth, is likely composed of silicate perovskite, a mineral with a structure wildly different from the minerals found in Earth's upper mantle and crust.

Mexican chemist Mario Molina and US chemist F Sherwood Rowland warn that the chlorofluorocarbons (CFCs) used in fridges and as aerosol propellants may be damaging the atmosphere's ozone layer, which filters out much of the Sun's ultraviolet radiation.

The Global Atmospheric Research Program (GARP) is launched. An international project, its aim is to provide a greater understanding of the mechanisms of the world's weather by using satellites and by developing a mathematical model of Earth's atmosphere.

1975

Five new nations, the USSR, West Germany, France, Japan, and the United Kingdom join the Deep Sea Drilling Project (DSDP) to form the International Phase of Ocean Drilling (IPOD).

Mineral physicists David Mao and Peter Bell of the Geophysical Laboratory in Washington, DC, use a diamond-anvil cell to produce pressures exceeding a million atmospheres.

22 January

The USA launches *Landsat 2*; it is positioned 180° from *Landsat 1* (launched 23 July 1972). The two together provide regular images of the Earth, including the capability to provide a view of the same geographical area of Earth with the same Sun angle every nine days, which is important in monitoring changes on the surface of the Earth.

16 October

The USA launches the first *Geostationary Operational Environmental Satellite* (GOES); it provides 24-hour coverage of US weather.

1976–1979

The International Magnetosphere Study conducts a three-year observation of the Earth's magnetosphere and its effects on the lower stratosphere including the disruptive effects of magnetic storms on communications.

1976

The American Panel on Atmospheric Chemistry warns that the Earth's ozone layer may be being destroyed by chloroflurocarbons (CFCs) from spray cans and refrigeration systems.

4 May

The US launches *Lageos* (Laser Geodynamic Satellite); it uses laser beams to make precise measurements of the Earth's movements in an attempt to improve the prediction of earthquakes. Placed in an orbit 9,321 km/ 5,793 mi high, it is expected to remain in orbit for 8 million years.

1977

Scientists from the project FAMOUS (French-American Mid-Ocean Undersea Study) in their deep-sea submersible vehicle *ALVIN* discover a host of strange life forms, such as large red and white tube worms, near undersea hot springs heated by ocean-ridge volcanism. The discovery proves the existence of life in extreme conditions.

1978

Core samples from the seabed are collected by the US research vessel *Glomar Challenger* from a record depth of 7,042 m/23,104 ft.

23 January

Sweden bans aerosol sprays because of their damaging effect on the environment. It is the first country to do so.

27 June

The US satellite *Seasat 1* is launched to measure the temperature of sea surfaces, wind and wave movements, ocean currents, and icebergs; it operates for 99 days before its power fails.

1979

3 June

Pemex Oil's offshore oil-well *Ixtoc 1* blows up, releasing an estimated 3 million barrels of crude oil into the Gulf of Mexico. The largest oil spill ever recorded, the slick spreads 965 km/600 mi to Texas, contaminating Gulf fisheries and beaches. The well defies capping efforts and it continues to disgorge oil until 24 March 1980.

1980

A ten-year World Climate Research Programme is launched to study and predict climate changes and human influence on climate change.

A thin layer of iridium-rich clay, about 65 million years old, is found around the world. US physicist Luis Alvarez suggests that it was caused by the impact of a large asteroid or comet which threw enough dust into the sky to obscure the Sun and cause the extinction of the dinosaurs.

North American geologists P Coney, D L Jones, and J W H Monger describe the North American Cordilleran orogen (in the Western USA and Canada) as a composite of 'suspect terranes', resulting in a new perspective on the construction of orogenic belts (in which mountain belts eventually form).

The US *Magsat* satellite completes its mapping of the Earth's magnetic field.

1981

The US Committee on the Atmosphere and Biosphere reports evidence linking acid rain to sulphur emissions from power plants.

The US government-commissioned *Global 2000 Report to the President* is published; it predicts global environmental catastrophe if pollution, industrial expansion, and population are not brought under control.

1982

The Convention on Conservation of Antarctic Marine Living Organisms comes into effect, establishing a protective oceanic zone around the continent.

1983

An exceptionally warm El Niño (warm water current) off the coasts of Ecuador and Peru drives the huge schools of anchovies, which thrive in the cold water, further offshore, resulting in the deaths of millions of larger fish and the birds which feed on them and the serious disruption of commercial fishing.

Soviet engineers drill a borehole to a depth of 12.3 km/7.6 mi at Zapolarny in the Kola peninsula, USSR – the deepest ever drilled.

Studies from the US *Lageos* satellite (launched 4 May 1976 to monitor slight crustal movements to help predict earthquakes) indicate that the Earth's gravitational field is changing.

The deep-sea drilling research ship *Glomar Challenger*, deployed by the Deep Sea Drilling Project (DSDP), is retired. It travelled a total of 375,632 nautical miles and acquired 19,119 cores during its 15-year service.

The Ocean Drilling Program (ODP) is established as the successor to the Deep Sea Drilling Project.

The skull of a creature called *Pakicetus* is discovered in Pakistan; estimated to be 50 million years old, it is intermediate in evolution between whales and land animals.

US chemist Mark Thiemens and his colleagues demonstrate that the production of ozone in the upper atmosphere causes separation of the two different heavy isotopes of oxygen independent of their masses. The discovery of this non-mass dependent kinetic partitioning of oxygen isotopes provides a new means of tracing the mixing of gases between Earth's different layers of atmosphere.

28 March
The first *TIROS N* satellite is launched by the USA as its contribution to the international Global Atmospheric Research Program (GARP), which continues with launches of Satellites by other countries later. The aim of the international project is to provide a greater understanding of the mechanisms of the world's weather by using satellites and by developing a mathematical model of the Earth's atmosphere.

21 July
Vostok station, Antarctica, records a temperature of −89.2°C/−128.6°F – the lowest on record.

1984

Australian geologists Bob Pidgeon and Simon Wilde discover zircon crystals in the Jack Hills north of Perth, Australia, that are estimated to be 4.276 million years old – the oldest rocks ever discovered.

1985

A well-preserved amphibian skeleton dated 340 million years old is discovered in Scottish oil shale. It is the earliest amphibian found.

The research drilling ship *SEDCO/BP 471* begins service as part of the new Ocean Drilling Program (ODP), replacing the retired *Glomar Challenger*. The ship is later renamed the *JOIDES Resolution*

1987

At a conference in Montreal, Canada, an international agreement, the Montreal Protocol, is reached to limit the use of ozone-depleting chlorofluorocarbons (CFCs) by 50% by the end of the century; the agreement is later condemned by environmentalists as 'too little, too late'.

US researchers prove that thunderstorm systems can propel pollutants into the lower stratosphere when they observe high levels of carbon monoxide and nitric acid at high altitude during a thunderstorm.

1988
A United Nations Environment Programme (UNEP) report claims that two-thirds of the world's urban population breathe 'disturbingly high levels' of sulphur dioxide and dust.

1989
The warmest year on record worldwide; environmentalists suggest this is due to the 'greenhouse effect'.

24 March
The *Exxon Valdez* oil tanker runs aground in Prince William Sound, Alaska, spilling an estimated 40,504,000 l/8,910,880 gal of oil. It is the largest oil spill in US history. Over 4,800 km/3,000 mi of shoreline are polluted.

27 March
As the oil spill from the tanker *Exxon Valdez* spreads over 100 sq mi/260 sq km, a state of emergency is declared in the area affected.

19 December
Nearly 19 million l/4.2 million gal of crude oil spill into the Atlantic from the Iranian tanker *Khark 5* following an explosion and fire.

1990
Canadian scientists discover fossils of the oldest known multicellular animals, dating from 600 million years ago.

27 February
The Exxon Corporation is indicted on five criminal charges relating to the 1989 Alaskan oil spill.

1991
'Biosphere 2', an experiment that attempts to reproduce the world's biosphere in miniature within a sealed glasshouse, is launched in Arizona, USA. Eight people remain sealed inside for two years.

A borehole in the Kola Peninsula in Arctic Russia, begun in the 1970s, reaches a record depth of 12,261 m/40,240 ft.

A circular impact structure of Cretaceous–Tertiary (K–T) age is found buried beneath in Mexico's Yucatan peninsula. Called the Chicxulub crater, it is the best candidate for the K–T impact site envisioned by Alvarez and others.

Less than 50% of the world's rainforest remains.

The World Ocean Experiment (WOCE) programme is set up to monitor ocean temperatures, circulation, and other parameters.

24 January
Iraq begins to pump Kuwaiti oil into the Persian Gulf during the Gulf War, creating the world's largest oil spill. About 6–8 million barrels of oil are spilled, polluting 675 km/420 mi of coastline.

4 April
The US Environmental Protection Agency announces ozone layer depletion at twice the speed previously predicted.

17 July
The European Space Agency's first remote-sensing satellite (*ERS-1*) is launched into polar orbit to monitor the Earth's temperature from space.

1992

3–14 June
The United Nations Conference on Environment and Development is held in Rio de Janeiro, Brazil. It is attended by delegates from 178 countries, most of whom sign binding conventions to combat global warming and to preserve biodiversity (the latter is not signed by the USA).

1993

An ice core drilled in Greenland, providing evidence of climate change over 250,000 years, suggests that sudden fluctuations have been common and that the recent stable climate is unusual.

The US scientist Albert Bradley develops the Autonomous Benthic Explorer (ABE), a robotic submersible that can descend to depths of 6.4 km/4 mi and remain at such depths for up to one year.

1994

The United Nations (UN) Basel Convention bans the transport of hazardous waste, from the 25 industrialized nations that make up the Organization for Economic Cooperation and Development (OECD), across international boundaries.

14 June
Representatives of 25 European countries and Canada sign a United Nations protocol in Oslo, Norway, to reduce sulphur emissions, a cause of acid rain.

4 July
Electrical flashes known as 'Sprites' – upper atmosphere optical phenomena associated with thunderstorms – are first examined by plane, by a team from the University of Alaska Statewide System, Fairbanks, Alaska, USA.

1995

At the international climate conference held in Melbourne, Australia, it is reported that periodic disruptions of surface currents (which may cause climate changes) have been discovered in the Atlantic and Indian Oceans.

The Prince Gustav Ice Shelf and the northern Larsen Ice Shelf in Antarctica begin to disintegrate – a result of global warming.

US and French geophysicists discover that the Indo-Australian plate split in two in the middle of the Indian Ocean about 8 million years ago.

April
The European Space Agency's Earth-sensing satellite ERS-2 is launched successfully. It will work in tandem with ERS-1, launched in 1991, to take measurements of global ozone.

August
The US Environmental Protection Agency approves the sale of genetically modified maize, which contains a gene from a soil bacterium that produces a toxin fatal to the European corn borer – a pest that causes approximately $1 billion damages annually.

1996

Scientists from the Scott Polar Institute, using data from the European Space Agency's ERS-1 satellite, discover a 14,000-sq-km/5,400-sq-mi, 125-m/

410-ft-deep lake, 4 km/2.5 mi under the Antarctic ice sheet. Called Lake Vostok after the Russian ice-drilling station it lies beneath, the ice sheet, which acts as a blanket, and a pressure of 300–400 atmospheres allow the water to remain liquid.

US geophysicists discover that the Earth's core spins slightly faster than the rest of the planet.

2 June
US scientists at the National Oceanic and Atmospheric Administration in Washington, DC, announce the first decline in levels of ozone-depleting chemicals in the air.

1997
February
The US zoologists Bill Detrich and Kirk Malloy show that the increased ultraviolet radiation caused by the hole in the ozone layer above Antarctica kills large numbers of fish in the Southern Ocean. Because their transparent eggs and larvae stay near the surface for up to a year, they are exposed to the full force of the ultraviolet rays. It is the first time ozone depletion in the Antarctic has been shown to harm organisms larger than one-celled marine plants.

26 March
The German ecologist Venugopalan Ittekkot shows that dams on the River Danube keep back silicate sediments and thus starve the Black Sea of food for many marine plants, and create ideal conditions for the growth of toxic competitors, altering the sea's ecosystem. Silicate-loving sea grasses have been replaced by nitrate-loving species such as dinoflagellates which cause poisonous red tides. It raises concerns that the world's thousands of dams may be slowly killing the world's seas.

11 June
French meteorologist Cyril Moulin shows that up to a billion tonnes of dust a year are blown off the arid drought-prone lands surrounding the Sahara Desert in north Africa and carried as far as the UK and the Caribbean. The amount has more than doubled in the past 30 years.

26 June
The second Earth Summit takes place in New York. Delegates report on progress since the 1992 Rio Summit and note that progress on the Rio biodiversity convention has been slower than on the convention on climate.

The delegates fail to agree on a deal to address the world's escalating environmental crisis. Dramatic falls in aid to the so-called Third World countries, which the 1992 summit promised to increase, are at the heart of the breakdown.

24 July
The Canadian researcher Richard Bottomley and colleagues date the 100-km/62-mi-wide Popigai impact crater in Siberia, thought to be the fifth-largest impact crater on Earth, to 35.7 million years old. They suggest that the meteorite that created it may be responsible for the mass extinction that occurred at the end of the Eocene and the start of the Oligocene geological periods.

25 July
The US researcher Joseph L Kirschvink and colleagues, by examining the record of remnant magnetism in very ancient rocks, discover that the outer layers of the Earth shifted by 90° relative to the core between about 535 and 520 million years ago. This major reorganization of the continents they suggest may have led to the Cambrian Explosion – the rapid appearance of abundant fossils in the geological record in the Cambrian Period, which began 540 million years ago.

4 August
Using computer models, British meteorologist Alan O'Neill demonstrates a connection between the collapse of anchovy fishing in Peru, drought in Australia, and the late arrival of India's monsoons and El Niño, the warm water current off South America's west coast.

7 September
The Australian researcher William de la Mare, using old whaling records which record data on every whale caught since the 1930s, including the ship's latitude, announces the discovery that Antarctic sea-ice could have decreased by up to a quarter between the mid-1950s and the 1970s. The finding has major implications, both for global climate conditions as well as for whaling.

1998
8 April
The International Union for the Conservation of Nature (IUCN), based in Switzerland, publishes a survey which reports that one in every eight known plant species in the world is in danger of becoming extinct.

Chronology

17 April
An iceberg 40 km/25 mi long and 4.8 km/3 mi wide breaks off from the Larson B ice shelf in Antarctica. Global warming is thought to be the cause.

October
The size of the hole in the ozone layer is measured as three times the size of the USA, bigger than it has ever been before. Its cause may be the adverse effects of El Niño on the climate.

2 November
Britain's Meteorological Office predicts that large areas of the Amazon rainforest will begin to die in around 2050, releasing carbon into the atmosphere and causing global warming to accelerate faster than previously forecasted.

2–14 November
Delegates from more than 160 countries attend a summit in Buenos Aires, Argentina, to discuss implementing the 1997 Kyoto Protocol to reduce global warming. Participants agree to devise strategies by 2000 for reducing emissions of greenhouse gases.

17 December
The United Nations World Meteorological Organization reports that 1998 is the warmest year ever, with a global average temperature of 14.4°C/58°F. Many scientists blame the greenhouse effect, caused by the burning of industrial gases.

1999

23 June
According to the 1999 World Disasters Report, published by the Red Cross, refugees from environmental disasters such as drought, floods, and deforestation totalled 25 million in 1998, outnumbering war refugees for the first time.

17 November
Scientists from the University of Washington in Seattle publish data showing that the Arctic icecap has shrunk by around 40% over the past 50 years, probably due to global warming. The Canadian Wildlife Service reports that polar bears in the area are in danger of starving to death because of their shortened hunting season.

3 Biographical Sketches

Adams, Leason Heberling (1887–1969)

US geophysicist. Adams published a new method of heating and gradually cooling (or 'annealing') optical glass in 1920. Adams gained further recognition for his research into the elastic properties of minerals and rocks at high pressure, especially as they related to the composition of the Earth.

Adams was born in Cherry Vale, Kansas, and was at the Geophysical Laboratory of Carnegie Institution of Washington 1910–52. He served as Professor of Geophysics at the University of California: Los Angeles from 1958 until his retirement in 1965.

Anderson, Don Lynn (1933–)

US geophysicist and seismologist. Anderson became director of the Seismology Laboratory at the California Institute of Technology (Caltech) in 1967. His research is focused upon theoretical seismology and the evolution, structure, and composition of the Earth and planets. He has edited the periodical *Physics of the Earth and Planetary Interiors*. He was awarded the 1998 National Medal of Science.

Anderson was born in Maryland. He received his BS in Geology and Geophysics from Rensselaer Polytechnic Institute in 1955 and his PhD in Geophysics and Mathematics from Caltech in 1962. He was Director of the Seismological Laboratory of the California Institute of Technology from 1967–89 before becoming professor of geophysics in the Division of Geological and Planetary Sciences at Caltech.

Anthes, Richard Allen (1944–)

US meteorologist. Anthes' work on computer modelling of the atmosphere improved meteorologists' ability to understand climate and predict weather events. In 1988, he became president of the University Corporation for Atmospheric Research, umbrella organization of US National Center for Atmospheric Research.

Anthes was born in St Louis, Missouri, and educated at the University of Wisconsin-Madison. He joined the US Weather Bureau in 1962.

Bailey, Edward Batersby (1881–1965)

British geologist who was noted for investigating the geological structure of Scotland. In the course of his work, he discovered and described many new phenomena.

Bailey was on the staff of the Geological Survey and Museum, 1902–29, and was appointed its director, 1937–45. As professor of geology at the University of Glasgow, 1929–37, he specialized in studying Dalradian and Tertiary igneous rocks, and was the first to describe cauldron subsidences and first recumbent folds (or 'nappes') in the Scottish Highlands. He was made a Fellow of the Royal Society in 1930.

Benioff, (Victor) Hugo (1899–1968)

US seismologist and inventor. Benioff devised and developed many seismic detection instruments, including a seismograph for measuring travel-time curves of earthquake waves that became the basis for the Geneva Conference nuclear detection system. He was the first to describe the plane of earthquake foci extending from Pacific Ocean trenches to the Earth's shallower mantle, now known as the Benioff Zone.

Birch, (Albert) Francis (1903–)

US geophysicist. Birch's studies of elasticity of rocks and minerals at high pressures and temperatures contributed extensively to knowledge of the composition of the Earth. His proposal that the Earth's innermost core is not liquid, but consists mainly of solid crystalline iron, was confirmed in 1971. Birch was born in Chevy Chase, Maryland, and educated at Harvard University. He became emeritus professor of geology at Harvard University.

Bjerknes, Vilhelm Firman Koren (1862–1951)

Norwegian scientist whose theory of polar fronts formed the basis of all modern weather forecasting and meteorological studies. He also developed hydrodynamic models of the oceans and the atmosphere and showed how weather prediction could be carried out on a statistical basis, dependent on the use of mathematical models.

Bjerknes was professor at Stockholm, Sweden, and Leipzig, Germany, before returning to Norway and founding the Bergen Geophysical Institute in 1917.

During World War I, Bjerknes instituted a network of weather stations throughout Norway; coordination of the findings from such stations led him and his co-workers to the theory of polar fronts, based on the discovery that the atmosphere is made up of discrete air masses displaying dissimilar features.

He coined the word 'front' to denote the boundary between such air masses.

Boltwood, Bertram (1870–1927)

US radiochemist who pioneered the use of radioactive elements as tools for dating rocks and determined the age of the Earth to be at least 2.2 billion years old. He discovered ionium and also made the first observations of the phenomenon of isotopy.

By studying abundances of radioactive elements in ores, Boltwood deduced that the radium present in an ore was the product of the breakdown of uranium in the ore and that uranium ultimately would decay to lead. In 1907, he demonstrated that by knowing the rate at which uranium decays (its half-life) he could calculate the age of a mineral by measuring the relative proportions of its uranium and lead. Boltwood dated rocks from several localities using his uranium–lead technique, obtaining ages between 410 million to 2.2 billion years old. His efforts showed that the Earth was significantly older than previously thought.

Boltwood was born in Amherst, Massachusetts, and studied chemistry at Yale from 1889 to 1992. He went on to study at Munich and Leipzig before returning to Yale as a professor 1897–1900. In 1900, he left Yale to work with geologist J H Platt. It was during this time that Ernest Rutherford and Frederick Soddy discovered that radioactive elements, such as thorium and uranium, turn into other elements by radioactive decay.

Boltwood returned to Yale in 1907 as a professor of radiochemistry. He went on to discover ionium. In his efforts to separate ionium from thorium he made the first observations of isotopes, since ionium is not a separate element, but actually an isotope of thorium.

Bowen, Norman Levi (1887–1956)

Canadian geologist whose work helped found modern petrology (the study of rocks). He demonstrated the principles governing the formation of magma by partial melting, and the fractional crystallization of magma.

Born in Kingston, Ontario, Bowen was educated at the local Queen's University, then moved to the recently founded Geophysical Laboratory in Washington, DC, USA. His findings on the experimental melting and crystallization behaviour of silicates and similar mineral substances were published from 1912 onwards. In *The Evolution of Igneous Rocks* (1928) he dealt particularly with magma, becoming known as the head of the 'magmatist school' of Canadian geology.

Bridgman, Percy Williams (1882–1961)

US physicist. His research into machinery producing high pressure led in 1955 to the creation of synthetic diamonds by General Electric. He was awarded the Nobel Prize for Physics in 1946 for his development of high-pressure physics.

Born in Cambridge, Massachusetts, he was educated at Harvard, where he spent his entire academic career.

Bridgman's experimental work on static high pressure began in 1908, and because this field of research had not been explored before, he had to invent much of his own equipment; for example, a seal in which the pressure in the gasket always exceeds that in the pressurized fluid. The result is that the closure is self-sealing. His discoveries included new, high-pressure forms of ice.

His technique for synthesizing diamonds was used to synthesize many more minerals and a new school of geology developed, based on experimental work at high pressure and temperature. Because the pressures and temperatures that Bridgman achieved simulated those deep below the ground, his discoveries gave an insight into the geophysical processes that take place within the Earth. His book *Physics of High Pressure* 1931 still remains a basic work.

Bullard, Edward Crisp (1907–1980)

English geophysicist who, with US geologist Maurice Ewing, founded the discipline of marine geophysics. He pioneered the application of the seismic method to study the sea floor. He also studied continental drift before the theory became generally accepted. Knighted 1953.

Bullard was born in Norwich and educated at Cambridge. During World War II he did military research, and he continued to advise the UK Ministry of Defence for several years after the war. He was professor of geophysics at the University of Toronto, Canada 1948–50; director of the National Physical Laboratory at Teddington, Middlesex 1950–57; and head of geodesy and geophysics at Cambridge 1957–74. He was also a professor at the University of California from 1963, and advised the US government on nuclear-waste disposal.

Bullard's earliest work was to devise a technique (involving timing the swings of an invariant pendulum) to measure minute gravitational variations in the East African Rift Valley. He then investigated the rate of efflux (outflow) of the Earth's interior heat through the land surface; later he devised apparatus for measuring the flow of heat through the deep sea floor.

While at Toronto University, Bullard developed his 'dynamo' theory of geomagnetism, according to which the Earth's magnetic field results from convective movements of molten material within the Earth's core.

Biographical Sketches

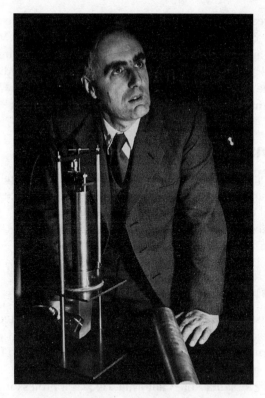

English geophysicist Edward Bullard director of the National Physical Laboratory. In 1941, at the age of 34, he was one of the youngest Fellows ever elevated to the Royal Society. Hulton-Deutsch Collection/Corbis

Charney, Jule Gregory (1917–1981)

US meteorologist. Renowned for his use of computers to generate forecasts (numerical weather), his mathematical models to describe weather and climate, and his theories of the instability of atmospheric pressure. In Princeton, New Jersey between 1948–56, he and others pioneered the first computer-generated weather forecast using the ENIAC (Electronic Numerical Integrator and Calculator).

He was born in San Francisco, the son of Russian immigrants. Following his doctorate, he studied with the famous Carl-Gustaf Rossby before travelling to the University of Oslo in Norway as a National Research Fellow. This success prompted his establishing a Joint Numerical Weather Prediction Unit in Maryland, which generated daily predictions of gross climate and weather patterns. At the Massachusetts Institute of Technology (1956–81) he chaired the Committee on International Meteorological Cooperation and helped organize the Global Atmospheric Research Program; their findings are expected to significantly advance understanding of the atmosphere.

He is remembered as the 'father of modern meteorology'.

Cloos, Ernst (1898–1974)

German geologist. His research concerned the use of small-scale deformation structures in the elucidation of major tectonic forces, and the use of scale model experiments simulating these effects.

He was professor at Johns Hopkins University, Baltimore, USA, from 1941 until his death.

Cox, Allan V(erne) (1926–1987)

US geophysicist. His palaeomagnetic research substantiated evidence of periodic reversals in the Earth's magnetic field, confirmed plate tectonic theories of continental drift and seafloor spreading, and led to publication of a new geological time scale in 1982.

He was born in Santa Ana, California. He was a professor of geophysics at Stanford University in 1967, then dean of their School of Earth Sciences in 1979.

Crutzen, Paul (1933–)

Dutch meteorologist who shared the Nobel Prize for Chemistry in 1995 with Mexican chemist Mario Molina and US chemist F Sherwood Rowland for their work in atmospheric chemistry, particularly concerning the formation and decomposition of ozone. They explained the chemical reactions which are destroying the ozone layer.

Crutzen, while working at Stockholm University in 1970, discovered that the nitrogen oxides NO and NO_2 speed up the breakdown of atmospheric ozone into molecular oxygen. These gases are produced in the atmosphere from nitrous oxide N_2O which is released by micro-organisms in the soil. He showed that this process is the main natural method of ozone breakdown. Crutzen also discovered that ozone-depleting chemical reactions occur on the surface of cloud particles in the stratosphere.

Crutzen was born in Amsterdam. He received his doctor's degree in meteorology from Stockholm University in 1973. He is currently at the Max Planck Institute for Chemistry in Mainz, Germany.

Dziewonski, Adam Marian (1936–)

Polish-born seismologist. A member of the Center for Earth and Planetary Physics at Harvard University, he became chairman of Harvard's Department of Geological Science. He utilized seismic wave movement and geomagnetic soundings to determine physical properties of the Earth's crust.

He was born in Lwów and he had performed seismological research in Poland before joining the Southwest Center for Advanced Studies at the University of Texas in 1965.

Du Toit, Alexander Logie (1878–1948)

South African geologist. His work was to form one of the foundations for the synthesis of continental drift theory and plate tectonics that created the geological revolution of the 1960s.

The theory of continental drift put forward by German geophysicist Alfred Wegener inspired Du Toit's book *A Geological Comparison of South America and South Africa* (1927), in which he suggested that they had probably once been joined. In *Our Wandering Continents* (1937), he maintained that the southern continents had, in earlier times, formed the supercontinent of Gondwanaland, which was distinct from the northern supercontinent of Laurasia.

Du Toit was born near Cape Town and studied there, then went to the UK and studied at Glasgow and the Royal College of Science, London. He spent 1903–20 mapping for the Geological Commission of the Cape of Good Hope.

Edinger, Tilly (1897–1967)

German-born US palaeontologist, born Johanna Gabrielle Ottilie Edinger. Her work in vertebrate palaeontology laid the foundations for the study of palaeoneurology when she demonstrated that the evolution of the brain could be studied directly from fossil cranial casts.

Her research shed new light on the evolution of the brain and showed that the progression of brain structure does not proceed at a constant rate in a given family but varies over time; also that the enlarged forebrain evolved several times independently among advanced groups of mammals and there was no single evolutionary scale.

Edinger was born in Frankfurt and studied there and at Heidelberg and Munich. With the Nazis's rise to power, she was forced to leave Germany. After a year in the UK, she went to Cambridge, Massachusetts, in 1940, to take up a job at the Museum of Comparative Zoology at Harvard. Edinger's main works are *Die fossilen Gehirne/Fossil Brains* (1929) and *The Evolution of the Horse Brain* (1948).

German-born US palaeontologist Tilly Edinger in 1950. Edinger was a leading figure in the field of 20th-century vertebrate palaeontology and laid the foundations for the study of palaeoneurology. Bettmann/Corbis

Elsasser, Walter Maurice (1904–1991)

German-born US geophysicist. He pioneered analysis of the Earth's former magnetic fields, which are frozen in rocks. His research in the 1940s yielded the dynamo model of the Earth's magnetic field. The field is explained in terms of the activity of electric currents flowing in the Earth's fluid metallic outer core. The theory premises that these currents are magnified through mechanical motions, rather as currents are sustained in power-station generators.

Born in Mannheim and educated at Göttingen, Elsasser left in 1933 following Hitler's rise to power, and spent three years in Paris working on the theory of atomic nuclei. After settling in 1936 in the USA and joining the staff of the California Institute of Technology, he specialized in geophysics. Elsasser became a professor at the University of Pennsylvania in 1947; in 1962 he was made professor of geophysics at Princeton.

Eskola, Pentti Eelis (1883–1964)

Finnish geologist. He was one of the first to apply physicochemical postulates on a far-reaching basis to the study of metamorphism, thereby laying the foundations of most subsequent studies in metamorphic petrology.

Throughout his life Eskola was fascinated by the study of metamorphic rocks, taking early interest in the Precambrian rocks of England. Building largely on Scandinavian studies, he was concerned to define the changing pressure and temperature conditions under which metamorphic rocks were formed. His approach enabled comparison of rocks of widely differing compositions in respect of the pressure and temperature under which they had originated.

Eskola was born in Lellainen and educated as a chemist at the University of Helsinki before specializing in petrology. In the early 1920s he worked in Norway and in Washington, DC, USA. He was professor at Helsinki 1924–53.

Ewing, (William) Maurice (1906–1974)

US geologist. His studies of the ocean floor provided crucial data for the plate tectonics revolution in geology in the 1960s. He demonstrated that midocean ridges, with deep central canyons, are common to all oceans.

Using marine sound-fixing and ranging seismic techniques and pioneering deep-ocean photography and sampling, Ewing ascertained that the crust of the Earth under the ocean is much thinner (5–8 km/3–5 mi thick) than the continental shell (about 40 km/25 mi thick). His studies of ocean sediment showed that its depth increases with distance from the midocean ridge, which gave clear support for the hypothesis of seafloor spreading.

Ewing was born in Lockney, Texas, and studied at the Rice Institute in Houston. He developed his geological interests by working for oil companies. In 1944 he joined the Lamont–Doherty Geological Observatory, New York. From 1947 he was professor of geology at Columbia University, while also holding a position at the Woods Hole Oceanographic Institute.

Fleming, John Adam (1877–1956)

US geophysicist. Long associated with the Department of Terrestrial Magnetism of the Carnegie Institution of Washington, DC, he designed several geomagnetic observatories. He made important inventions in the field of geomagnetism and contributed to research in solar and lunar physics.

He was born in Cincinnati, Ohio. He joined the Department of Terrestrial Magnetism of the Carnegie Institution of Washington, DC, became its director, and continued to serve as an adviser in international scientific relations after his retirement. During his directorship, he designed geomagnetic observatories in Huancayo, Peru, Wateroo, Australia, and Kensington, Maryland. He took charge of the Institution's World War II contracts dealing with radio communications, ordnance devices, magnetic instruments, and ionospheric research to advance knowledge of the Earth's magnetic field. He was one of the first officers of the American Geophysical Union, serving as general secretary and editor of their periodical. He expanded the field of geomagnetism by inventing or modifying both terrestrial and oceanographic geomagnetic instruments, and designing innovative isomagnetic world charts.

Gardner, Julia (Anna) (1882–1960)

US geologist and palaeontologist. Her work was important for petroleum geologists establishing standard stratigraphic sections for Tertiary rocks in the southern Caribbean.

Her work on the Cenozoic stratigraphic palaeontology of the Coastal Plain, Texas, and the Rio Grande Embayment in northeast Mexico led to the publication of *Correlation of the Cenozoic Formations of the Atlantic and Gulf Coastal Plain and the Caribbean Region* (1943) (with two coauthors).

Julia Gardner was born in South Dakota and educated at Bryn Mawr College and Johns Hopkins University. She worked for the US Geological Survey 1911–54. During World War II, she joined the Military Geologic Unit where she helped to locate Japanese beaches from which incendiary bombs were being launched, by identifying shells in the sand ballast of the balloons.

Goldring, Winifred (1888–1971)

US palaeontologist. Her research focused on Devonian fossils and during the late 1920s and the 1930s, as well as geologically mapping the Coxsackie and Berne quadrangles of New York, she developed and maintained the State Museum's public programme in palaeontology.

Goldring was born near Albany, New York State, and educated at Wellesley College. She worked 1914–54 at the New York State Museum. Goldring began her work in palaeobotany in 1916. In 1939 she was made state palaeontologist. Her works include *The Devonian Crinoids of the State of New York* (1923) and *Handbook of Palaeontology for Beginners and Amateurs* (1929–31). She did much to popularize geology.

Goldschmidt, Victor Moritz (1888–1947)

Swiss-born Norwegian chemist. He did fundamental work in geochemistry, particularly on the distribution of elements in the Earth's crust. He considered the colossal chemical processes of geological time to be interpretable in terms of the laws of chemical equilibrium.

Using X-ray crystallography Goldschmidt was able to show that, given an electrical balance between positive and negative ions, the most important factor in crystal structure is ionic size. Exhaustive analysis of results from geochemistry, astrophysics, and nuclear physics led to his work on the cosmic abundance of the elements and the links between isotopic stability and abundance. Studies of terrestrial abundance reveal about eight predominant elements. Recalculation of atom and volume percentages lead to the remarkable notion that the Earth's crust is composed largely of oxygen anions (90% of the volume), with silicon and the common metals filling up the rest of the space.

Goldschmidt was born in Zürich but moved to Norway as a child and studied at the University of Christiania (now Oslo). He was professor and director of the Mineralogical Institute 1914–29, when he moved to Göttingen, Germany. The rise of Nazism forced him to return to Norway in 1935, but during World War II he had to flee again, first to Sweden and then to Britain, where he worked at Aberdeen and Rothamsted (on soil science). He returned to Norway after the end of the war.

During World War II Goldschmidt carried a cyanide capsule for suicide should the Germans have invaded Britain. When a colleague asked for one, he was told: 'Cyanide is for chemists; you, being a professor of mechanical engineering, will have to use the rope.'

Gould, Stephen Jay (1941–)

US palaeontologist and writer. In 1972 he proposed the theory of punctuated equilibrium, suggesting that the evolution of species did not occur at a steady rate but could suddenly accelerate, with rapid change occurring over a few hundred thousand years. His books include *Ever Since Darwin* (1977), *The Panda's Thumb* (1980), *The Flamingo's Smile* (1985), and *Wonderful Life* (1990).

Gould was born in New York and studied at Antioch College, Ohio, and Columbia University. He became professor of geology at Harvard 1973 and was later also given posts in the departments of zoology and the history of science.

Gould has written extensively on several aspects of evolutionary science, in both professional and popular books. His *Ontogeny and Phylogeny* (1977) provided a detailed scholarly analysis of his work on the developmental process of recapitulation. In *Wonderful Life* he drew attention to the diversity of the fossil finds in the Burgess Shale Site in Yoho National Park, Canada, which he interprets as evidence of parallel early evolutionary trends extinguished by chance rather than natural selection.

Gutenberg, Beno (1889–1960)

German seismologist. During his graduate studies at the University of Göttingen, Germany, he made the first known correct determination of the size and composition of the Earth's inner core. Gutenberg was born in Darmstadt, Germany. He left Germany in 1930 to join the seismology laboratory at the California Institute of Technology, where he collaborated with Charles Richter to develop the definitive scale of earthquake magnitude, then continued research on Earth structure, seismic waves, and stratospheric temperatures.

Hess, Harry Hammond (1906–1969)

US geologist who in 1962 proposed the notion of seafloor spreading. This played a key part in the acceptance of plate tectonics as an explanation of how the Earth's crust is formed and moves.

Hess was born in New York and studied at Yale and Princeton, where he eventually became professor. From 1931, he carried out geophysical research into the oceans, continuing during World War II while in the navy. Later he was one of the main advocates of the Mohole project, whose aim was to drill down through the Earth's crust to gain access to the upper mantle.

Building on the recognition that certain parts of the ocean floor were anomalously young, and the discovery of the global distribution of midocean ridges and central rift valleys, Hess suggested that convection within the Earth was continually creating new ocean floor, rising at midocean ridges and then flowing horizontally to form new oceanic crust. It would follow that the further from the midocean ridge, the older would be the crust – an expectation confirmed by research in 1963.

Hess envisaged that the process of seafloor spreading would continue as far as the continental margins, where the oceanic crust would slide down beneath the lighter continental crust into a subduction zone, the entire operation thus constituting a kind of terrestrial conveyor belt.

Holmes, Arthur (1890–1965)

English geologist who helped develop interest in the theory of continental drift. He also pioneered the use of radioactive decay methods for rock dating, giving the first reliable estimate of the age of the Earth.

Holmes was born in Newcastle-upon-Tyne and studied at Imperial College, London. He was appointed head of the Geology Department at Durham in 1924, moving to Edinburgh University in 1943.

Holmes was convinced that painstaking analysis of the proportions in rock samples of elements formed by radioactive decay, combined with a knowledge of the rates of decay of their parent elements, would yield an absolute age. From 1913 he used the uranium–lead technique systematically to date fossils whose relative (stratigraphical) ages were established but not the absolute age.

In 1928, Holmes proposed that convection currents within the Earth's mantle, driven by radioactive heat, might furnish the mechanism for the continental drift theory broached a few years earlier by German geophysicist Alfred Wegener. In Holmes's view, new rocks were forming throughout the ocean ridges. Little attention was given to these ideas until the 1950s.

His books include *The Age of the Earth* (1913), *Petrographic Methods and Calculations* (1921), and *Principles of Physical Geology* (1944).

Knopf, Eleanora Frances (1883–1974)

US geologist, born Eleanora Frances Bliss, who studied metamorphic rocks. She introduced the technique of petrofabrics to the USA.

Bliss was born in Rosemont, Pennsylvania, and studied at Bryn Mawr College and the University of California at Berkeley. She spent most of her career working for the US Geological Survey. During the 1930s she was also a visiting lecturer at Yale and at Harvard.

In 1913 in Pennsylvania, she discovered the mineral glaucophane, previously unsighted in the USA east of the Pacific. In the 1920s Knopf studied the Pennsylvania and Maryland piedmont and the geologically complex mountain region along the New York–Connecticut border.

The technique of petrofabrics had been developed in Austria at Innsbruck University. Knopf applied it to the study of metamorphic rocks and wrote about it in *Structural Petrology* (1938).

Le Pichon, Xavier (1937–)

French geophysicist who worked out the motions of the Earth's six major lithospheric plates. His work was instrumental in the development of plate tectonics.

In 1968 he published *Sea-floor Spreading and Continental Drift* in which he depicted the Earth's lithosphere divided into six major plates. The boundaries between the plates were shown to have high seismic activity and occurred along mid-ocean ridges, island arcs, active orogenic (mountain-building) belts and the transform faults revealed earlier by J Tuzo Wilson.

Le Pichon was born in Vietnam. He became a research assistant at Columbia University's Lamont–Doherty Geological Observatory, New York, in 1963. He received many awards, including the Maurice Ewing Medal of the American Geophysical Union (1984) and the Wollaston Medal of the Geological Society of London (1991). He is currently a professor in the Department de Geologie at the Ecole Normale Superieure in Paris, France.

Biographical Sketches

Libby, Willard Frank (1908–1980)

US chemist who was awarded the Nobel Prize for Chemistry in 1960 for his development in 1947 of radiocarbon dating as a means of determining the age of organic or fossilized material.

Libby was born in Grand Valley, Colorado, and studied at the University of California, Berkeley. During World War II he worked on the development of the atomic bomb (the Manhattan Project). In 1945 he became professor at the University of Chicago's Institute for Nuclear Studies. He was a member of the US Atomic Energy Commission 1954–59, and then became director of the Institute of Geophysics at the University of California.

Having worked on the separation of uranium isotopes for producing fissionable uranium-238 for the atomic bomb, he turned his attention to carbon-14, a radioactive isotope that occurs in the tissues of all plants and animals, decaying at a steady rate after their death. He and his co-workers accurately dated ancient Egyptian relics by measuring the amount of radiocarbon they contained, using a Geiger counter. By 1947 they had developed the technique so that it could date objects up to 50,000 years old.

US chemist Willard Libby in 1954 when he became a member of the US Atomic Energy Commission, nominated by President Eisenhower. Corbis/Bettmann-UPI

Lovelock, James Ephraim (1919–)

British scientist who began the study of CFCs in the atmosphere in the 1960s, and who later elaborated the 'Gaia hypothesis' – the concept of the Earth as a single organism, or ecosystem. The Gaia hypothesis, named after an ancient Greek earth goddess, views the planet as a self-regulating system in which all the individual elements coexist in a symbiotic relationship. In developing this theory (first published in 1968), Lovelock realized that the damage effected by humans on many of the Earth's ecosystems was posing a threat to the viability of the planet itself.

Lovelock invented the electron capture detector in the 1950s, a device for measuring minute traces of atmospheric gases. He developed the Gaia hypothesis while researching the possibility of life on Mars for NASA's space programme; it was not named 'Gaia' until some years later, at the suggestion of the writer William Golding.

Matsuyama, Motonori (1884–1958)

Japanese geophysicist who determined that the Earth's magnetic field reverses its polarity periodically throughout its geological history. He also

English scientist James Lovelock, who developed the concept of the Earth as a single organism, in what we now know as the Gaia hypothesis. George W Wright/Corbis

pioneered the use of gravimetry in finding geological structures below the Earth's surface.

In 1930, Matsuyama began studying magnetism preserved in rocks. It was known as early as 1906 that some rocks were magnetized such that they pointed toward magnetic north, while the magnetism of other rocks pointed in the opposite direction. Matsuyama studied these magnetic anomalies and suggested they were due to a reversal in the polarity of the Earth's magnetic field. The Matsuyama Epoch, a major polar reversal occuring approximately 0.5–2.5 million years ago, is named after him.

Matsuyama was born in Usa, Japan, and studied mathematics and physics at Hiroshima Normal College and Imperial University, Kyoto. He became a professor at Imperial University in 1913 where he studied gravity determination using pendulum techniques. He carried out extensive gravity surveys in Manchuria, Korea, and in the Japan Trench.

Miller, Stanley Lloyd (1930–)

US chemist. In the early 1950s, under laboratory conditions, he tried to recreate the formation of life on Earth. To water under a gas mixture of methane, ammonia, and hydrogen, he added an electrical discharge. After a week he found that amino acids, the ingredients of protein, had been formed.

Miller was born in Oakland, California, and studied at the uni-versities of California and Chicago. From 1960 he held appointments at the University of California in San Diego, rising to professor of chemistry.

Miller made his experiment while working for his PhD under Harold Urey, using the components that had been proposed for the Earth's primitive atmosphere by Urey and Russian biochemist Alexandr Oparin. The electrical discharge simulated the likely type of energy source.

Mohorovičić, Andrija (1857–1936)

Croatian seismologist and meteorologist who discovered the Mohorovičić discontinuity, the boundary between the Earth's crust and the mantle.

In 1909 after a strong earthquake occurred in the Kulpa Valley south of Zagreb, Mohorovičić discovered two distinct sets of P- and S-waves (types of seismic wave) – one set arriving earlier than the other. He deduced that one set of waves was slower than the other because it had travelled through denser material. Mohorovičić proposed that the Earth's surface consists of an outer layer of rocky material approximately 30 km/19 mi thick, which overlies a denser mantle. Later research has shown that the boundary between these two layers, the 'Moho', lies at a depth of 5–10 km/3–6 mi beneath the ocean crust and approximately 35 km/21 mi beneath the crust

US chemist Stanley Miller holding the reaction chamber in which methane, ammonia, water vapour, and hydrogen gases (the 'atmosphere') were exposed to an electrical current to simulate lightning in his famous 1953 experiment that showed how the first life forms might have arisen. Jim Sugar Photography/Corbis

of the continents. Seismic waves travel nearly 20% slower below the Moho than above and it is regarded as the bottom of the Earth's crust.

Born in Volosko, Croatia, Mohorovičić studied mathematics and physics at Prague University. After seven years as a school teacher, he became a professor at the Zagreb Technical School and later Zagreb University. In 1892 he became director of the Meteorological Station in Zagreb, which later became the Royal Regional Centre for Meteorology and Geodynamics, establishing a seismological observatory there in 1901.

The Royal Regional Centre for Meteorology and Geodynamics was renamed the Geophysical Institute in 1921. Mohorovičić continued seismological research there until 1926.

Oldham, Richard Dixon (1858–1936)

Irish seismologist who discovered the Earth's core and first distinguished between primary and secondary seismic waves.

During his tenure at the Geological Survey of India (1879–1903), Oldham investigated an earthquake in Assam in June 1897 and recognized that there were two phases of seismic waves recorded by the seismograph, primary waves (P-waves) and secondary waves (S-waves), and that these should propagate through the whole Earth.

In 1906, while analyzing seismic records, Oldham noticed an area on the globe in which P-waves were not detected. Every time an earthquake occurred, this P-wave 'shadow zone' appeared on the opposite side of the globe. Oldham demonstrated that the Earth had a core that was causing the primary waves to refract (bend) away, leaving a seismic shadow. Oldham further realized that the material of the core was significantly different from the rest of the Earth, since the strength of the secondary waves was greatly reduced, creating another shadow zone. In 1919, after the theory of seismic waves had been developed by Wiechert, Oldham suggested that the core may be liquid.

Oldham was born in Dublin, Ireland, and was educated at the Royal School of Mines.

Patterson, Colin (1933–1998)

English palaeontologist whose work on fossil classification by evolutionary relationship using paraphyletics (the creation of taxonomic groups that include some, but not all, of the descendants of a common ancestor) was influential in reforming the fossil record.

Patterson was born in Hammersmith, London, and attended university at Imperial College, London. He spent most of his working life at the London Natural History Museum.

Richter, Charles Francis (1900–1985)

US seismologist, deviser of the Richter scale used to indicate the strength of the waves from earthquakes. He was born near Hamilton, Ohio.

Stommel, Henry M(elson) (1920–)

US oceanographer. He was considered the leading authority on Gulf Stream dynamics. Using physical models, he developed the first theory of the Atlantic current, which he expanded into his definitive book, *The Gulf Stream* (1955). Working with various colleagues, he made major contributions to studies of cumulus clouds, oceanic salinity and thermal gradients, and plankton distribution.

He was born in Wilmington, Delaware. He was a physical oceanographer at the Woods Hole Oceanographic Institution 1944–59, before joining the Massachusetts Institute of Technology (MIT) 1959–60. He left MIT to teach at Harvard 1960–63, returned to MIT 1963–78, then moved back to Woods Hole as senior scientist in 1979. From 1954–76, Stommel established numerous long-term international projects to gather data on the world's ocean currents and geochemistry of the sea. In recognition of his profound influence on geophysical hydrodynamics and climatology, he shared Sweden's prestigious Crafoord Prize in 1983.

Urey, Harold Clayton (1893–1981)

US chemist. In 1932 he isolated heavy water and was awarded the Nobel Prize for Chemistry in 1934 for his discovery of deuterium (heavy hydrogen).

During World War II he was a member of the Manhattan Project, which produced the atomic bomb, and after the war he worked on tritium

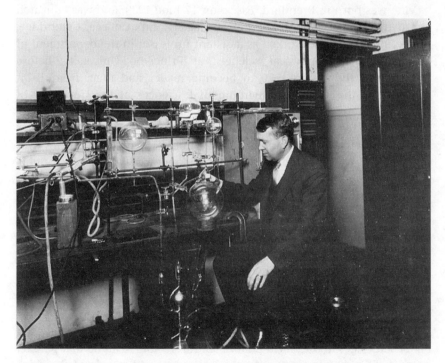

US chemist Harold Urey in 1933 in the new laboratory that had just been built for him at Columbia University, expressly for the purpose of producing heavy water.
Corbis/Bettmann-UPI

(another isotope of hydrogen, of mass 3) for use in the hydrogen bomb, but later he advocated nuclear disarmament and world government.

Urey was born in Indiana and educated at Montana State University. He became professor of chemistry at Columbia 1934, and was at Chicago 1945–58.

After deuterium, Urey went on to isolate heavy isotopes of carbon, nitrogen, oxygen, and sulphur. His group provided the basic information for the separation of the fissionable isotope uranium-235 from the much more common uranium-238.

Urey also developed theories about the formation of the Earth. He thought that the Earth had not been molten at the time when its materials accumulated. In 1952, he suggested that molecules found in its primitive atmosphere could have united spontaneously to give rise to life. The Moon, he believed, had a separate origin from the Earth.

Vine, Frederick John (1939–1988)

English geophysicist whose work was an important contribution to the development of the theory of plate tectonics.

Vine was born in Brentford, Essex, on 17 June 1939, and was educated at Latymer Upper School, London and at St John's College, Cambridge. From 1967 to 1970 he worked as assistant professor in the department of geological and geophysical sciences at Princeton University, before returning to the UK in 1970 to become reader and then, from 1974, professor in the school of environmental sciences at the University of East Anglia.

In Cambridge in 1963, Vine collaborated with his supervisor Drummond Hoyle Matthews (1931–   ), and wrote a paper, 'Magnetic anomalies over ocean ridges', which provided additional evidence for Harry Hess's seafloor-spreading hypothesis of 1962. Alfred Wegener's original 1912 theory of continental drift (Wegener's hypothesis) had been met with hostility at the time because he could not explain why the continents had drifted apart, but Hess had continued his work developing the theory of seafloor spreading to explain the fact that as the oceans grew wider, the continents drifted apart.

Following the work of Brunhes and Motonori Matuyama in the 1920s on magnetic reversals, Vine and Matthews predicted that new rock emerging from the oceanic ridges would intermittently produce material of opposing magnetic polarity to the old rock. They applied palaeomagnetic studies to the ocean ridges in the North Atlantic and were able to argue that parallel belts of different magnetic polarities existed on either side of the ridge crests. This evidence was vital proof of Hess's hypothesis. Studies on ridges in other oceans also showed the existence of these magnetic anomalies.

Vine and Matthews's hypothesis was widely accepted in 1966 and was confirmation of Hess's earlier work. Their work was crucial to the development of the theory of plate tectonics and revolutionized the earth sciences.

Wegener, Alfred Lothar (1880–1930)

German meteorologist and geophysicist whose theory of continental drift, expounded in *Origin of Continents and Oceans* (1915), was originally known as 'Wegener's hypothesis'. His ideas can now be explained in terms of plate tectonics, the idea that the Earth's crust consists of a number of plates, all moving with respect to one another.

Wegener was born in Berlin and studied at Heidelberg, Innsbruck, and Berlin. From 1924 he was professor of meteorology and geophysics at Graz, Austria. He completed three expeditions to Greenland and died on a fourth.

Wegener supposed that a united supercontinent, Pangaea, had existed in the Mesozoic. This had developed numerous fractures and had drifted

German meteorologist and geophysicist Alfred Wegener in 1926, when he was rector of the Handel School in Germany. Corbis/Bettmann-UPI

apart some 200 million years ago. During the Cretaceous, South America and Africa had largely been split, but not until the end of the Quaternary had North America and Europe finally separated; the same was true of the break between South America and Antarctica. Australia had been severed from Antarctica during the Eocene.

Wilson, John Tuzo (1908–1993)

Canadian geologist and geophysicist who established and brought about a general understanding of the concept of plate tectonics.

Born in Ottawa, Wilson studied geology and physics – an original combination that led directly to the development of the science of geophysics – at the University of Toronto, and obtained his doctoral degree at Princeton University, New Jersey, USA. His particular interest was the movement of the continents across the Earth's surface – then a poorly understood and not widely accepted concept known as 'continental drift'. He spent 28 years as professor of geophysics at the University of Toronto, retiring in 1974 just as interest in plate tectonics was developing worldwide. From then on he was the director-general of the Ontario Science Centre and later the chancellor of York University, Toronto, finally retiring 1987. In 1957 he was the president of the International Union of Geodesy and Geophysics – the most senior administrative post in the field.

Wilson's great strength was in education. He pioneered hands-on interactive museum exhibits, and could explain complex subjects like the movement of continents, the spreading of ocean floors, and the creation of island chains by using astonishingly simple models. He was an active outdoor man, leading expeditions into the remote north of Canada, and he made the first ascent of Mount Hague in Montana, USA, 1935.

The Wilson Range in Antarctica is named after him.

Part Two

Directory of Organizations and Institutions	81
Selected Works for Further Reading	91
Web Sites	99
Glossary	123

4 Directory of Organizations and Institutions

American Association of Petroleum Geologists (AAPG)
International organization devoted to advancing the geological sciences as they relate to natural resources, including petroleum, natural gas, and minerals. The principal journal of the AAPG is the *AAPG Bulletin*. They also publish the *AAPG Explorer*, a news magazine, as well as numerous monographs.

Address
1444 South Boulder Avenue
PO Box 979, Tulsa
OK 74101, USA
phone: +1 (918) 584-2555
fax: +1 (918) 560-2652
Web site: http://www.aapg.org/indexaapg.html

American Geological Institute
Federation of 31 geoscience societies, including the American Geophysical Union (AGU), the American Association of Petroleum Geologists (AAPG), the Geological Society of America (GSA), and many more. Its goal is to provide a united voice for geoscientists in matters of societal concern. AGI publishes a monthly news magazine, *Geotimes*, as well as reference materials such as the *AGI Glossary of Geology* and the GeoRef database for geoscience literature.

Address
4220 King Street, Alexandria
VA 22302-1502, USA
Web site: http://www.agiweb.org/

American Geophysical Union (AGU)
International professional organization for scientists engaged in studies of Earth's oceans, atmospheres, crust, mantle, and core as well as other planets and space. AGU publishes the widely-read weekly newspaper *EOS*. It publishes eight scientific journals, including *Geophysical Research Letters*,

Journal of Geophysical Research, Reviews of Geophysics, Tectonics, Global Biogeochemical Cycles, Nonlinear Processes in Geophysics (copublished with the European Geophysical Society), *Water Resources Research*, and *Paleoceanography*. In addition, the organization has begun an electronic journal called *Earth Interactions*. There are two well-attended annual AGU meetings each year, the autumn meeting in San Francisco and the spring meeting at varied locations in eastern USA.

Address
2000 Florida Avenue NW
Washington
DC 20009-1277, USA
phone: +1 (202) 462-6900 and 1-800-966-2481 (toll-free in North America)
fax: +1 (202) 328-0566
e-mail: service@agu.org
Web site: http://www.agu.org/

American Institute of Hydrology

National organization dedicated to the certification and education of professionals in the hydrological sciences.

Address
2499 Rice St., Ste. 135
St. Paul
MN 55113-3724, USA
phone: +1 (612) 484-8169
fax: +1 (612) 484-8357
e-mail: aihydro@aihydro.org
Web site: http://www.aihydro.org/

American Institute of Professional Geologists

International association that certifies professional geologists and promotes their interests, lobbying on their behalf at the State and Federal level and working with regulatory boards and agencies.

Address
7828 Vance Drive, #103 Arvada
CO 80003-2125, USA
phone: +1 (303) 431-0831
fax: +1 (303) 431-1332
e-mail: aipg@aipg.org
Web site: http://www.aipg.org/Scriptcontent/index.cfm

Association of Engineering Geologists
National organization promoting the application of geological techniques to construction and engineering. AEG has several publications, including the *Environmental & Engineering Geoscience Journal*.

Address
323 Boston Post Road, Suite 2D
Sudbury
Massachusetts 01776, USA
phone: +1 (508) 443-4639
fax: +1 (508) 443-2948
e-mail: aeghq@aegweb.org
Web site: http//www.aegweb.org

Association for Exploration Geochemists
For those interested in the field of natural resource exploration and applied geochemistry.

Address
PO Box 26099, 72 Robertson Road
Napean, Ontario, Canada K2H 9RO
phone: +1 (613) 828 0199
fax: +1 (613) 828 9288
Web site: http://www.aeg.org

Association for Women Geoscientists
A society for the professional development of female geoscientists and the promotion of geosciences to girls and young women.

Address
PO Box 280, Broomfield
CO 80038–0280, USA
Web site: http://www.awg.org

Geochemical Society
Academic society for research into the geochemistry of Earth, Moon, and meteorites.

Address
Department of Earth and Planetary Sciences
Washington University, 1 Brookings Drive
St Louis
MO 63130–4899, USA
phone: +1 (314) 935 4131
fax: +1 (314) 935 4121
Web site: http://gs.wustl.edu

Geological Association of Canada

A national society for geoscientists whose membership includes scientists from across Canada and around the world.

Address
Department of Earth Sciences
Alexander Murray Bldg (Rm ER4063)
Memorial University of Newfoundland
St John's, NF, Canada A1B 3X5
phone: +1 (709) 737 7660
fax: +1 (709) 737 2532
e-mail: gac@sparky2.esd.mun.ca
Web site: http://sparky2.esd.mun.ca/~gac/

Geological Society of America (GSA)

American professional organization for earth scientists from all sectors, including academia, government, and industry. The GSA caters to the geological sciences and publishes two major journals, *Geology* and the *GSA Bulletin*. In addition to the popular annual meeting of the GSA, there are regional meetings within the USA. The Society also sponsors several Penrose Conferences each year, with the aim of fostering free exchange of scientific ideas on specified topics.

Address
PO Box 9140, Boulder
CO 80301-9140, USA
phone: +1 (303) 447-2020
fax: +1 (303) 447-1133
Web site: http://www.geosociety.org/

Geological Society (UK)

Society of UK professional geologists.

Address
Burlington House, Piccadilly
London W1V 9AG, UK
phone: +44 (020) 7434 9944
fax: +44 (020) 7439 8975
e-mail: enquiries@geolsoc.org.uk
Web site: http://www.geolsoc.org.uk/template.cfm?name=geohome

Geologists' Association
UK society for professionals and amateurs; includes a children's division.

Address
Burlington House, Piccadilly
London W1V 9AG, UK
phone: +44 (0171) 434 9298
fax: +44 (0171) 287 0280
e-mail: geol.assoc@btinternet.com
Web site: http://www.geologist.demon.co.uk/index.htm

Geoscience Information Society
Society of librarians, scientists, educators, and editors promoting the exchange of information in the geosciences. The Society publishes two major reference books: *The Union List of Geologic Field Trip Guidebooks of North America* (6th ed., 1996) and the *Directory of Geosciences Libraries, US and Canada* (5th ed., 1997).

Address
Geoscience Information Society
c/o American Geological Institute
4220 King Street, Alexandria
Virginia 22302, USA
Web site: http://www.geoinfo.org

Lunar and Planetary Institute
A focus for academic participation in the studies of the current state, evolution, and formation of the Solar System.

Address
3600 Bay Area Boulevard
Houston
TX 77058–1113, USA
Web site: http://www.lpi.usra.edu/lpi.html

Marine Geology & Geophysics Division of the NOAA National Geophysical Data Center
NGDC's Marine Geology & Geophysics Division and the co-located World Data Center for Marine Geology & Geophysics compiles and maintains extensive databases in both coastal and open ocean areas.

Address
E/GC3 325 Broadway
Boulder, CO USA 80305-3328
fax: +1 (303) 497 6513
Web site: www.ngdc.noaa.gov/mgg/mggd.html

Mineralogical Society of America

International organization devoted to advances in mineralogy, crystallography, and petrology founded in 1919. The Society publishes the journal *American Mineralogist* and a topical monograph series called *Reviews in Mineralogy*, which is popular worldwide. They also publish a newsletter *The Lattice*.

Address
1015 Eighteenth Street NW, Suite 601
Washington, DC 20036, USA
phone: +1 (202) 775-4344
fax: +1 (202) 775-0018
Web site: http://www.minsocam.org/

National Geophysical Data Center

Part of the US Department of Commerce's National Oceanic & Atmospheric Administration (NOAA) that coordinates and manages environmental data from its five centres: marine geology and geophysics, palaeoclimatology, solar-terrestrial physics, solid earth geophysics, and glaciology (snow and ice). As well as fostering research in these areas, the centre aims to make data from its 300-plus databases accessible to a variety of users.

Address
National Geophysical Data Center
NOAA, Mail Code E/GC, 325 Broadway
Boulder
CO 80303-3328, USA
phone: +1 (303) 497-6826
fax: +1 (303) 497-6513
e-mail: info@ngdc.noaa.gov
Web site: http://www.ngdc.noaa.gov/mgg/mggd.html

National Weather Service

Provides weather, hydrologic, and climate forecasts and warnings for the USA. The National Weather Service is part of the NOAA (National Oceanic and Atmospheric Administration).

Address
National Weather Service, NOAA
1325 East–West Highway, Silver Spring
MD 20910, USA
Web site: http://www.nws.noaa.gov

Paleontological Research Institution

Institution and museum affiliated with Cornell University that houses extensive fossil collections – as well as laboratories, offices, and a 50,000 volume research library – and fosters research in palaeontology. The Institute publishes *Bulletins of American Paleontology*, 'the oldest continuously published palaeontological journal in the western hemisphere and one of the oldest in the world'.

Address
1259 Trumansburg Road, Ithaca
New York 14850, USA
phone: +1 (607) 273-6623
fax: +1 (607) 273-6620
Web site: http://www.englib.cornell.edu/pri/pri1.html

Paleontological Society

Society founded in 1908 to promote advances in palaeontology. Its members include professional scientists, land-managers, museum specialists, and technicians, as well as amateur collectors. The Society publishes the bi-monthly *Journal of Paleontology* and *Paleobiology*, which is published quarterly.

Address
Paleontological Society, Inc.
PO Box 1897
Lawrence
KS 66044-8897, USA

Seismological Society of America

Organization of geologists, geophysicists, seismologists, engineers, and policy-makers promoting developments in earthquake science. It publishes the *Seismological Research Letters* and the *Bulletin of the Seismological Society of America*, 'the premier journal of research in earthquake seismology and related disciplines since 1911'.

Address
201 Plaza Professional Building
El Cerrito
CA 94530, USA
phone: +1 (510) 525-5474
fax: +1 (510) 525-7204
e-mail: info@seismosoc.org
Web site: http://www.seismosoc.org

SEPM (Society for Sedimentary Geology)
Society for geologists specializing in sedimentary rocks and their environs.

Address
1741 E 71st Street
Tulsa
OK 74136-5108, USA
phone: +1 (918) 493-3361 or (800) 865-9765 (North America)
fax: +1 (918) 493-2093
e-mail: ngeeslin@sepm.org
Website: http://www.sepm.org/sepm.html

Society for Mining, Metallurgy, and Exploration, Inc
International organization established in 1857 to promote mining and minerals exploration. Its publications include *Mining Engineering* (monthly) and *Minerals and Metallurgical Processing* (quarterly).

Address
8307 Shaffer Parkway
PO Box 625002, Littleton
CO 80162-5002, USA
phone: +1 (303) 973-9550 x210
fax: +1 (303) 973-3845

Society for Organic Petrology
Organization devoted to studies of the origin of coal and kerogen, and organic geochemistry in general.

Address
Cortland Eble, TSOP Membership Chair
Kentucky Geological Survey
228 Mining and Minerals Building
University of Kentucky, Lexington
KY 40506, USA
phone: +1 (606) 257-5500
fax: +1 (606) 258-1049

Soil Science Society of America

Promotes advancements in soil research. Its publications include the *Soil Science Society of America Journal,* the *Journal of Environmental Quality,* and the *Journal of Production Agriculture.* **Date established** 1936

Address
677 South Segoe Road
Madison
WI 53711, USA
phone: +1 (608) 273 8095
fax: +1 (608) 273 2021
e-mail: headquarters@soils.org
Web site: http://www.soils.org

5 Selected Works for Further Reading

Alexander, D *Natural Disasters*, 1993

> A complete account of the natural, social, and political causes of natural disasters (including volcanic eruptions, earthquakes, and floods) and the possible responses.

Anderson, D L *Theory of the Earth*, 1989

> A high-level text, primarily centred on questions of the structure and chemistry of the interior of the Earth.

Barnes, J W *Basic Geological Mapping*, 1991

> An excellent pocket-sized field guide to mapping geological formations.

Barry, Roger G, and Chorley, Richard J *Atmosphere, Weather and Climate*, 1992

> A heavyweight volume for the dedicated reader. Everything is here from the gas content of the sky to the causes of climatic change.

Benton, M J *Basic Palaeontology*, 1997

> A great text on the anatomy and environment of major organisms found in the fossil record.

Bolt, B A *Earthquakes*, 1999

> A fascinating and well-written book by an expert in the field, for both professionals and interested laypeople. It details the physics and mechanics of earthquakes, how they affect Earth's surface, and how they are used to image the interior of the Earth.

Briggs, David *Encyclopedia of Earth Sciences*, 1992

> Solid and dependable textbook.

Broecker, W *How to Build a Habitable Planet*, 1985

> A complete, well-written, and easy to read book on the evolution of the earth.

Selected Works for Further Reading

Brown, G A and others *The Age of the Earth*, 1992

> The updated version of the Open University's textbook on earth science, bringing the original 1972 classic completely up to date.

Calder, Nigel *Restless Earth*, 1972

> This book represented something of a landmark in being one of the first general readers (it accompanied a television series) to examine the role of the newly forming concept of plate tectonics in accounting for the major physical features of the Earth's surface. It is superbly illustrated and is still highly regarded today.

Calder, Nigel *The Weather Machine*, 1974

> A very readable account of meteorological phenomena and events, by an expert in putting science into everyday language.

Carruthers, M W and Clinton, S *Pioneers of Geology: Discovering Earth's Secrets*, 2001

> Biographies of six of the most important geologists since the 18th century, including three from the 20th century: Alfred Wegener, Harry Hess, and Gene Shoemaker.

Cas, R A F *Volcanic Successions, Modern and Ancient*, 1987

> An interesting and well-illustrated text on volcanic rocks, interpreting history of volcanism through detailed study of the rocks.

Dana, J D *Dana's New Mineralogy: The System of Mineralogy of James Dwight Dana and Edward Salisbury Dana*, 1997

> The definitive source on minerals. Dana's mineralogy includes basic information on composition, structure, and provenance of every known mineral up to 1997.

Dasch, E Julius *Sedimentary Environments: Processes, Facies and Stratigraphy*, 1996

> Possibly the best guide to read the rocks.

Decker, R W and Decker, B B *Mountains of Fire: The Nature of Volcanoes*, 1991

> A well illustrated introduction to volcanoes, their rock types, and their effects.

Deer, W A *Rock-forming Minerals*, 1978

> An extremely useful book detailing the composition, structure, and environment of formation of all the major rock-forming minerals.

Selected Works for Further Reading

Dixon, Dougal *The Cambridge Encyclopedia of Earth Sciences*, 1992

> A useful coverage of the various aspects of geology and the techniques used to study them.

Drury, S A *Image Interpretation in Geology*, 1993

> A guide to techniques and applications in remote sensing.

Duff, P McL D *Holmes' Principles of Physical Geology*, 1992

> Update of Arthur Holmes' original textbook, used by geology students for more than 50 years.

Dunning, F W and others *The Story of the Earth*, 1981

> Gives an overview of the concept of plate tectonics and shows how volcanoes and earthquakes fit into the grand scheme.

Elsom, D *Planet Earth: the Making, Shaping and Workings of a Planet*, 1998

> A well-illustrated, well-written book for the general audience. Includes sections on the solid Earth, surficial processes, Earth's interior, weather and climate, and current environmental issues.

Emery, Dominic and Meyers, Keith *Sedimentary Environments: Processes, Facies and Stratigraphy*, 1996

> Excellent textbook devised from an in-house course run by BP.

Fortey, Richard *Fossils: The Key to the Past*, 1991

> Bringing fossils to life and applying this knowledge to geology and biology. An effortless browse through all the branches of the subject, introducing all the main fossil organisms at the same time.

Francis, P *Volcanoes: A Planetary Perspective*, 1993

> A well-written and well-illustrated book by a volcanologist, this book is intended for university students and professionals, but would also be interesting and informative to anyone interested in the details of volcanology.

Goldring, R *Field Palaeontology*, 1999

> An excellent text on conducting field investigations and on how to use palaeontological data.

Goudie, Andrew and Gardner, Rita *Discovering Landscape*, 1985

> This book aims to 'discover and try to explain some of the most appealing features of the natural landscape'. It is a splendid book for all those with an inquisitive mind who want to know a bit more about the history and

geology of well-known British sites such as Helvellyn, Lulworth Cove, and Cheddar Gorge.

Gould, Stephen Jay (ed) *The Book of Life*, 1993

The thinking person's palaeontological coffee table book (and why not? See Gould's panegyric on the significance of coffee tables and their books). First-class illustrations and chapters by an impressive cast list of other distinguished specialists with emphasis on vertebrates, including humans.

Greeley, R *Planetary Landscapes*, 1994

A fascinating book on the geology of the terrestrial planets, rocky satellites, and asteroids.

Greeley, R and Batson, R *NASA Atlas of the Solar System*, 1997

A beautifully illustrated, complete book on the astronomy, geology, and exploration of the Solar System.

Hall, A *Igneous Petrology*, 1996

A university text on the compositions, textures, and environments of formation of igneous rocks.

Hallam, A *A Revolution in the Earth Sciences*, 1973

An interesting and well-explained account of the scientific discoveries leading up to the theory of plate tectonics.

Hardy, Ralph; Wright, Peter; Gribbin, John; and Kington, John *The Weather Book*, 1982

A book to fascinate anyone with an interest in meteorology. Packed with colour pictures and amazing facts.

Houghton, J *Global Warming: The Complete Briefing*, 1997

The causes, effects, and politics of global warming, written by an expert in the field.

Jackson, J *Glossary of Geology*, 1997

The most comprehensive glossary around, and the one that professionals refer to. Published by the American Geological Institute.

Kaharl, Victoria A *Water Baby*, 1992

This is a fascinating account of the building and development of the research submersible *Alvin*. It is one of a handful of vehicles which takes scientists down into the depths to glimpse this alien realm. *Alvin* has been the platform from which researchers have made many startling discoveries. It is the oceanographic equivalent of the space shuttle.

Kearey, P and Vine, F *Global Tectonics*, 1996

An excellent, up-to-date text on plate tectonics, observation and theory.

Klein, C *Manual of Mineralogy (after James D Dana)*, 1993

A textbook of mineralogy, including the basics of mineral composition, structure, formation environments, and identification.

Manley, Gordon *Climate and the British Scene*, 1975

Originally one of the now collectable Collins New Naturalist Series, this classic account opens with a fascinating history of meteorological recording and, after detailed coverage of the climate of Britain, finishes with the impact of weather and climate on humans.

McGuire, B *Apocalypse*, 2000

A popular book on natural disasters, written by an expert in the field.

Milner, Angela *The Natural History Museum Book of Dinosaurs*, 1995

No book list in this subject can be without something on dinosaurs. This is an up-to-date look at their biology – what we can deduce about their lifestyles from modern research and new finds – together with the history and practice of dinosaur studies, from bones in the rock to a restoration of the whole animal.

Myers, Norman *The Gaia Atlas of Planet Management*, 1994

There are many atlases and reference books on the market today but there are few that are as informative and lavishly illustrated as this. It is intelligent and thought-provoking and there are many excellent thematic maps and colour photographs. It is divided into several sections including the land, the oceans, the elements, evolution, and civilization. I strongly recommend this book for all those with an interest and concern for the issues affecting the future of our planet.

Nolet, G *Seismic tomography*, 1987

About imaging the interior of the Earth with seismic waves.

Open University Oceanography course team *Waves, Tides and Shallow Water Processes*, 1989

First rate textbook intended for an oceanography home study course.

Selected Works for Further Reading

Open University, Oceanography course team *Ocean Basins: their Structure and Evolution*, 1989

First rate textbook intended for an oceanography home study course.

Pellant, Chris *The Practical Geologist*, 1990

It is normally not a good idea to base mineral identification on photographs – the diagnostic properties do not show. However, the photographs in this book are particularly good.

Press, Frank and Siver, Raymond *Understanding Earth*, 1998

An excellent introductory text.

Reading, Harold *Remote Sensing*, 1996

A comprehensive textbook that is also entertaining.

Ricciutti, E and Carruthers, M *Audubon First Field Guide: Rocks and Minerals*, 1998

A well-illustrated field guide, perfect for children or adults.

Rudwick, Martin *The Meaning of Fossils*, 1972

Widely regarded as the standard scholarly reference to the historical scientific issues and intellectual questions posed by palaeontology.

Sigurdsson, H *Melting the Earth: the History of Ideas on Volcanic Eruption*, 1999

A fascinating book on the history of volcanology.

Sigurdsson, H *Encyclopedia of Volcanoes*, 1999

The most up-to-date and complete resource on volcanology. A collection of papers on volcanology, from chemistry to hazard mitigation.

Simkin, Tom and Siebert, Lee *Volcanoes of the World: A Regional Directory, Gazetteer, and Chronology of Volcanism During the Last 10,000 Years*, 2000

A catalogue of all volcanoes active in the Holocene, with information on each of their recorded eruptions.

Skinner, B J *Blue Planet: An Introduction to Earth System Science*, 1995

A good introductory text on a reasonably simple level.

Smith, James *Introduction to Geodesy: the History and Concepts of Modern Geodesy*, 1998

A history of geodesy that is accessible to the general reader.

Stanley, S M *Earth and Life Through Time*, 1986

> Earth and life and the meaning of nearly everything. This is the American college solution to everything you wanted to know about fossils, evolution, and geology. Don't worry about the self-assessment questions unless you are a student – this book is here on its encyclopedic merit.

Summerfield, M A *Global Geomorphology: An Introduction to the Study of Landforms*, 1991

> An excellent, well-illustrated text on planetary landforms and the processes by which they evolve.

Summerfield, M A *Geomorphology and Global Tectonics*, 2000

> On the morphology of landforms of tectonic origin.

Swinnerton, H H *Fossils*, 1960

> A notable contribution to the classic *New Naturalist* series, so inevitably a very homely British bias. No matter – the range of fossil-bearing rocks in Britain is a good enough sample of the fossil record, which Swinnerton sets out with charm, enjoyment, and enthusiasm.

Tassy, Pascal *The Message of Fossils*, 1991, translated 1993

> An easy-going essay focused on the interface between palaeontology and evolution, which includes its more recent controversies.

Thurman, H V *Essentials of Oceanography*, 1999

> A basic university text on oceanography, well written and illustrated.

Trueman, A E *Geology and Scenery in England and Wales*, 1971

> For those with an interest in the geological development of particular landscapes in England and Wales such as the West Country Moors, the Cotswolds, or the Lake District, this is a must as it is both informative and highly readable.

van Rose, Susanna *Earthquakes*, 1983

> Describes the various phenomena associated with an earthquake, covers the known causes, and gives some case histories.

Waller, Geoffrey *Sealife: A Complete Guide to the Marine Environment*, 1996

> A splendidly comprehensive book covering evolution and biology of all groups of marine life, with attractive colour illustrations and a very useful practical section on sampling and recording techniques.

Selected Works for Further Reading

Wells, Sue, and Hanna, Nick *The Greenpeace Book of Coral Reefs*, 1992

The main thrust of this book is explaining the coral ecosystem and the complex cycles of life that are increasingly under threat by human activities.

Whittow, John *Disasters*, 1980

This fascinating book looks at the causes and effects of the major natural hazards such as earthquakes, landslides, and floods written by a very well-respected author. It contains some amazing and often chilling eye-witness accounts.

Wilhelms, D *To a Rocky Moon: A Geologist's History of Lunar Exploration*, 1994

A detailed account of the geological side of the US space programme, written by a geologist involved.

Wilson, M *Igneous Petrogenesis, a Global Tectonic Approach*, 1989

An excellent textbook discussing the details of the origin of igneous rocks.

Wilson, Edward O *The Diversity of Life*, 1992

An outstanding overview of the biodiversity crisis, as well as basic concepts such as evolutionary change, extinction, and speciation, written for the general public. The all-encompassing range, the compelling case for conservation, and the delightful natural history are virtues enough to recommend this book.

Windley, B F *Evolving Continents*, 1995

An upper level text on the origin and evolution of the continents.

6   Web Sites

Adventures in the Learning Web
http://www.usgs.gov/education/othered.html

Huge range of online educational geological resources, including sections on 'Finding your way with map and compass', 'Monitoring active volcanoes', and 'This dynamic Earth'. The site is run by the US Geological Survey, so some of the material is US-specific. However, a whole host of more general earth science topics are covered on this site in clear, well-illustrated sections. This site is regularly updated with information on any natural disasters as and when they happen.

Alternative Technology
http://www.environment.gov.au/education/aeen/pd/tsw/mod23/mod23.htm

Part of an electronic teacher's guide 'Teaching for a Sustainable World: International Edition', this site also contains useful information for individual students.

Arctic Circle
http://arcticcircle.uconn.edu/

Well-written site with information about all aspects of life in the Arctic. There are sections on history, natural resources, the rights of indigenous peoples, and issues of environmental concern.

Ask-A-Geologist
http://walrus.wr.usgs.gov/docs/ask-a-ge.html/

'Do you have a question about volcanoes, earthquakes, mountains, rocks, maps, ground water, lakes, or rivers?' If you do, the US Geological Survey invites you to ask one of their earth scientists a question at this site.

Ask an Energy Expert
http://www.eren.doe.gov/menus/energyex.html

The Energy Efficiency and Renewable Energy Clearinghouse (EREC) invites you to submit a question on 'energy-efficient and renewable energy technologies'.

Atlapedia Online
http://www.atlapedia.com/online/map_index_pol.htm

Database of maps which allows you to view a detailed political map for almost any country in the world. You can also choose to see a physical map for the same country. Also included on this site are detailed facts about each of the countries.

Avalanche!
http://www.pbs.org/wgbh/nova/avalanche/

Companion to the US Public Broadcasting Service (PBS) television programme *Nova*, this page follows an intrepid documentary team as they set out to film an avalanche. It provides information on the causes of avalanches, as well as details on how the film crew avoided getting swept away by them. There are six video clips of actual avalanches in progress and a number of still photos. You can download a transcript of the television programme.

Avalanche Awareness
http://www-nsidc.colorado.edu/NSIDC/EDUCATION/AVALANCHE/

Description of avalanches, what causes them, and how to minimize dangers if caught in one. There is advice on how to determine the stability of a snowpack, what to do if caught out, and how to locate people trapped under snow. Nobody skiing off piste should set off without reading this.

Before and After the Great Earthquake and Fire: Early Films of San Francisco, 1897–1916
http://lcweb2.loc.gov/ammem/papr/sfhome.html

Collection of 26 early films depicting San Francisco before and after the 1906 disaster, including a 1915 travelogue that shows scenes of the rebuilt city and the Panama Pacific Exposition, and a 1916 propaganda film. Should you not wish to download the entire film, each title contains sample still-frames. There is also background information about the earthquake and fire, and a selective bibliography.

Big Empty
http://www.blm.gov/education/great_basin/great_basin.html

Sponsored by the US Bureau of Land Management, this site explores the Great Basin of the western USA and its desert ecosystem of plants, animals, and minerals. This site includes information on the scarcity of water, modern and environmental challenges, mining, grazing, wild horses and burros, and a look at methods to preserve and rehabilitate the ecosystem. Maps and photographs complement the text.

Bodleian Library Map Room – The Map Case
http://www.rsl.ox.ac.uk/nnj/mapcase.htm

Broad selection of images from the historical map collection of the Bodleian library. Visitors can choose between rare maps of Oxfordshire, London, areas of Britain, New England, Canada, and more. The maps can be viewed by thumbnail and then selected in their full GIF or JPEG version.

British Geological Survey Global Seismology and Geomagnetism Group Earthquake Page
http://www.gsrg.nmh.ac.uk/gsrg.html

Fascinating maps showing the location and relative magnitude of recent UK earthquake activity make up just a small part of this Web site. Also included are historical and archive information, as well as descriptions of felt effects and hazards. Watch out if you live in Wolverhampton!

British Geological Survey
http://www.bgs.ac.uk/

Site of the UK's national centre for earth science information and its foremost supplier of geoscience solutions. It acquires and maintains up-to-date knowledge of the UK and its continental shelf by means of systematic geological, geophysical, geochemical, hydrogeological, and geotechnical surveys underpinned by high quality research.

Cambrian Period: Life Goes for a Spin
http://www.sciam.com/explorations/082597cambrian/powell.html

Part of a larger site maintained by *Scientific American,* this page reports on the research of Joseph Kirschvink of the California Institute of Technology which suggests that the so-called 'Cambrian Explosion' resulted from a sudden shifting of the Earth's crust. The text includes hyperlinks to further information, and there is also a list of related links, including one to figures, diagrams, and information from Kirschvink's research paper.

Cartographic Images Home Page
http://www.iag.net/~jsiebold/carto.html

Treasure trove of cartographic images spanning from 6,000 BC to 1700 AD, put into context with the help of individual or group monographs and a rich bibliography. A valuable and neatly designed resource of astonishing dimensions.

Web Sites

Causes of Climate Change
http://www.geog.ouc.bc.ca/physgeog/contents/7y.html

From a much larger site 'Fundamentals of Physical Geography' set up by a Canadian University and covering many aspects of the area of study. The site is clear and well organized and contains maps and diagrams.

Clouds and Precipitation
http://ww2010.atmos.uiuc.edu/(Gh)/guides/mtr/cld/home.rxml

Illustrated guide to how clouds form and to the various different types. The site contains plenty of images and a glossary of key terms in addition to further explanations of the various types of precipitation.

Clouds from Space
http://www.hawastsoc.org/solar/eng/cloud1.htm

This site offers a unique look at clouds, containing photographs of various cloud types taken from space including thunderstorms over Brazil, jet stream cirrus clouds, and a description of how clouds form.

Coastal Features and Processes
http://www.zephryus.demon.co.uk/geography/resources/revision/coast.html

This glossary of essential terms relating to coastal landscapes would make a helpful revision aid. The straightforward text layout and concise descriptions make this a good page to download and keep.

Coasts in Crisis
http://pubs.usgs.gov/circular/c1075/

Full text of a US Geological Survey publication. This detailed site uses intelligent text and photographs to show a coastline under siege. It describes the various types of coastline, methods of shoring them up, and the human and meteorological forces attacking the coast. This is an excellent resource for anyone studying coastal erosion, deposition and protection.

Composition of Rocks
http://www.geog.ouc.bc.ca/physgeog/contents/10d.html

From a much larger site 'Fundamentals of Physical Geography' set up by a Canadian University and covering many aspects of the area of study. The site is clear and well organized and contains maps and diagrams.

Coriolis Effect
http://www.physics.ohio-state.edu/~dvandom/Edu/coriolis.html

Subtitled 'a (fairly) simple explanation', this site contains a description of the principles involved, and is aimed at non-physicists.

Cracking the Ice Age
http://www.pbs.org/wgbh/nova/ice/

Companion to the US Public Broadcasting Service (PBS) television programme *Nova*, this page provides information about glaciation, the natural changes in climate over the past 60 million years, the greenhouse effect, global warming, and continental movement. There is also a list of related links.

Dan's Wild Wild Weather Page
http://www.whnt19.com/kidwx/index.html

Introduction to the weather for kids. It has pages dealing with everything from climate to lightning, from satellite forecasting to precipitation – all explained in a lively style with plenty of pictures.

Department of Atmospheric Sciences
http://www.atmos.uiuc.edu/

Masses of data and other information on climate around the planet. Among the site's more attractive features is the use of multimedia instructional modules, customized weather maps, and real-time weather data.

Dinofish.com
http://www.dinofish.com/

Contains much historical information about the discovery, study, and conservation of the coelacanth, a 400-million-year-old species of fish, as well as recent news and listings of scientific articles. Also on the site is a virtual swimming coelacanth, some online videos, and a shop for coelacananth enthusiasts!

Disasters: Panoramic Photographs, 1851–1991
http://lcweb2.loc.gov/cgi-bin/query/r?ammem/pan:@and(+fires+earthquake+storms+railroad+accidents+floods+cyclones+hurricane+))

Part of the Panoramic Photo Collection of the US Library of Congress, this page features 144 panoramic photographs of natural and human-caused disaster scenes, mostly in the USA, between 1851 and 1991. To narrow your search, click on New Search and enter a specific disaster or location. The images include brief notes. Click on the images to increase their size.

Donald L Blanchard's Earth Sciences' Web Site
http://webspinners.com/dlblanc/

Informative articles on palaeoclimate, plate tectonics, and palaeontology.

Double Whammy
http://www.sciam.com/explorations/1998/011998asteroid/

Part of a larger site maintained by *Scientific American*, this page explores the catastrophic impact that an asteroid's crashing into the sea would have on civilization and the environment. Animated simulations show the effects of impact and the effect that a tsunami would have on the eastern seaboard of the USA if an asteroid struck the Atlantic Ocean. Learn about the tsunami that struck Prince William Sound, Alaska, in 1964 after an underwater earthquake. The text includes hypertext links to further information and a list of related links.

Dust Bowl
http://www.usd.edu/anth/epa/dust.html

Historical information on the US mid-West dust bowl of the 1930s. The site features period photographs and an MPG video clip of a dust storm taken from original film footage (please beware the lengthy download time for this clip).

Earth and Heavens – The Art of Map Maker
http://portico.bl.uk/exhibitions/maps/

Excerpts from a major exhibition (now closed) at the British Library on the way maps and mapmaking have been used over the years to make statements about humankind's relationship to the world and the mysteries of the universe.

Earth Introduction
http://www.hawastsoc.org/solar/eng/earth.htm

Everything you ever wanted to know about planet Earth can be found at this site, which contains a table of statistics, photographs taken from space, radar-generated images of the planet, animations of the Earth rotating, and more.

Earthquakes and Plate Tectonics
http://wwwneic.cr.usgs.gov/neis/plate_tectonics/rift_man.html

US Geological Survey National Earthquake Information Centre site, explaining the relationship between plate tectonics and earthquakes.

Earth Sciences
http://dir.yahoo.com/science/earth_sciences/

Yahoo! directory of earth science, which provides links to hundreds of earth science sites, organized by discipline, and increasing in number every day.

Earth's Seasons: Equinoxes, Solstices, Perihelion, and Aphelion
http://aa.usno.navy.mil/AA/data/docs/EarthSeasons.html

Part of a larger site on astronomical data maintained by the US Naval Observatory, this site gives the dates and hours (in Universal Time) of the changing of the seasons from 1992 through to 2005. It also includes sections of 'Frequently Asked Questions' and research information.

Edwards Aquifer Home Page
http://www.edwardsaquifer.net/

Guide to the Edwards Aquifer (a rock formation containing water) in Texas – one of the most prolific artesian aquifers in the world.

El Niño Theme Page
http://www.pmel.noaa.gov/toga-tao/el-nino/nino-home.html

Wealth of scientific information about El Niño (a 'disruption of the ocean-atmosphere system in the tropical Pacific') with animated views of the monthly sea changes brought about by it, El Niño-related climate predictions, and forecasts from meteorological centres around the world. It also offers an illuminating 'Frequently Asked Questions' section with basic and more advanced questions as well as an interesting historical overview of the phenomenon starting from 1550.

EqIIS – Earthquake Image Information System
http://www.eerc.berkeley.edu/cgi-bin/eqiis_form?eq=4570&count=1

Fully searchable library of almost 8,000 images from more than 80 earthquakes. It is possible to search by earthquake, structure, photographer, and keyword.

Erosion and Deposition
http://www.geog.ouc.bc.ca/physgeog/contents/11g.html

Description of two important geological processes, part of a much larger site on physical geography, set up by a Canadian university. The hyperlinked text explains the three stages in the process of erosion – detachment, entrainment, and transport – as well as the causes of deposition.

Essential Guide To Rocks
http://www.bbc.co.uk/education/rocks/

Entertaining and informative site based on the BBC television series of the same name. This site shows in particular how geology and the earth sciences can be seen within cities, the home, and even the bathroom. Another feature is a series of 'virtual walks' showing rocks and minerals in surprising places in London, England, and other cities.

Exploring the Environment
http://www.cotf.edu/ete/main.html

Learning modules on environment-related earth science topics. Issues covered include biodiversity, land use, global warming, and water pollution. Information pages are accompanied by interactive learning activities.

Exploring the Tropics
http://www.mobot.org/MOBOT/education/tropics/

Missouri Botanical Garden guide to tropical rainforests covering issues such as 'Are all tropical forests, rain forests?', 'Effects of elevation on climate and vegetation', and 'Causes of destruction'.

Facing The Future: People and the Planet
http://www.facingthefuture.org/index4.htm

Guide to population issues, aimed at young people. The site tackles issues such as 'How many people can the Earth support?' and covers both personal and global solutions.

Features Produced by Running Water
http://www.zephryus.demon.co.uk/geography/resources/revision/river.html

Revision page aimed at GCSE geography students, explaining essential terms and concepts to do with rivers and running water. The layout is simple and quick to load, and the definitions are clear and concise.

Fernando De Noronha
http://www.noronha.com.br/indexe.htm

Examination of the unique ecosystem of the isolated archipelago of Fernando de Noronha, off the northeastern coast of Brazil. With dozens of colourful photographs, this site provides an informative insight into the islands and their history and includes a photo gallery of some of the flora and fauna. There are also details on how to get there and where to stay.

Flood!
http://www.pbs.org/wgbh/nova/flood/

Companion to the US Public Broadcasting Service (PBS) television programme *Nova*, this page concerns many aspects of flooding. It takes an historical look at floods and examines the measures that engineers have taken to combat them. Three major rivers are discussed: the Yellow, Nile, and Mississippi. In addition to learning about the negative effects of floods, you can also find out about the benefits that floods bestow on farmland. There are many images dispersed throughout the site, plus an audio clip of a flood in progress.

Geographia
http://www.geographia.com/

Excellent site, designed for travellers but containing a vast amount of information on countries all over the world. Five regional sections are accompanied by special features on selected locations. The text is supplemented by pictures, videos, and audio clips.

Geographical Study Resources
http://www.m8i.net/

Extensive geographical resource, which includes outline maps in GIF format, and geographical photographs, both of which are freely available to download. There is also a guide to writing coursework and essays.

Geologylink
http://www.geologylink.com/

Comprehensive information on geology featuring a daily update on current geologic events, virtual classroom tours, and virtual field trips to locations around the world. You will also find an in-depth look at a featured event, geologic news and reports, an image gallery, glossary, maps, and an area for asking geology professors your most perplexing questions, plus a list of references and links.

Glacier
http://www.glacier.rice.edu/

Find out about the coldest, windiest place on Earth, and the geological processes behind its formation. Full of detail about what it's really like to go on an expedition to Antarctica, this well-designed site has ample information on ice and glaciers, and explains why the ice-sheet that covers all but 2% of Antarctica has expanded and contracted throughout its history.

Web Sites

Glaciers
http://www-nsidc.colorado.edu/glaciers/

Comprehensive information about glaciers from the US National Snow and Ice Data Centre. There are explanations of why glaciers form, different kinds of glaciers, and what they may tell us about climate change. There are a number of interesting facts and a bibliography about the compacted tongues of ice which cover 10% of the land surface of our planet.

Global Climate Change Information Programme (GCCIP)
http://www.doc.mmu.ac.uk/aric/gcciphm.html

Established in October 1991, the GCCIP provides an information link between scientists (both natural and social), politicians, economists, and the general public, on the subjects of climate change and air quality. Their Web site includes a number of essays on related issues.

Global Drainage Basins Database
http://grid2.cr.usgs.gov/dem/basins.html

Informative United Nations Environment Programme (UNEP) database of drainage basins worldwide. This detailed, scientific site includes a good, clear explanation of what drainage basins are, and why they are important.

Greatest Places
http://www.greatestplaces.org/

This site takes you on a journey to seven of the most geographically dynamic locations on Earth. It features stylish presentation with extensive pictures and cultural commentary.

Great Globe Gallery
http://hum.amu.edu.pl/~zbzw/glob/glob1.htm

Over 200 globes and maps, showing the Earth from all angles, including space shots, political maps, the ancient world, geographical features, and animated spinning globes.

Ground Beneath
http://library.thinkquest.org/27026/

ThinkQuest project, particularly aimed at young people. This well-organized site is a guide to 'what's beneath the Earth's crust and all about plate tectonics'.

Groundwater Quality and the Use of Lawn and Garden Chemicals by Homeowners
http://www.ext.vt.edu/pubs/envirohort/426-059/426-059.html

> Extensive information about the problem of keeping groundwater free from garden chemicals. There is a description of groundwater, as well as detailed information on pesticides and their use. There are also notes on applying lawn and garden chemicals, and what should be done with the leftovers.

Hurricane & Tropical Storm Tracking
http://hurricane.terrapin.com/

> Follow the current paths of Pacific and mid-Atlantic hurricanes and tropical storms at this site. Java animations of storms in previous years can also be viewed, and data sets for these storms may be downloaded. Current satellite weather maps can be accessed for the USA and surrounding region.

Hydrologic Cycle
http://ww2010.atmos.uiuc.edu/(Gh)/guides/mtr/hyd/home.rxml

> Easy to understand and well-illustrated guide to the water cycle. The site is part of an online meteorology guide provided by the University of Illinois.

Hydrology Primer
http://wwwdutslc.wr.usgs.gov/infores/hydrology.primer.html

> Information from the US Geological Survey about all aspects of hydrology. The 'clickable' chapters include facts about surface water and groundwater, the work of hydrologists, and careers in hydrology. For answers to further questions click on 'ask a hydrologist', which provides links to other US national and regional sources.

Internal and External Friction
http://www.zephryus.demon.co.uk/geography/resources/fieldwork/fluvial/fric.html

> Page explaining the concept of friction in the context of rivers. Part of a much larger site containing resources for secondary school geography, this page describes how friction is crucial to the way in which a river is formed, and how it acts.

Web Sites

Introduction to Environmental Education
http://www.environment.gov.au/education/aeen/pd/tsw/module2/module2.htm

Part of an electronic teacher's guide 'Teaching for a Sustainable World: International Edition', this site also contains useful information for individual students.

Introduction to Plate Tectonics
http://www.hartrao.ac.za/geodesy/tectonics.html

Detailed guide to the background and major characteristics of plate tectonic theory. This site includes sections on fossil distribution, lithology, palaeomagnetism, oceanography, and 'What causes plates to move?'.

Introduction to the Ecosystem Concept
http://www.geog.ouc.bc.ca/physgeog/contents/9j.html

Clear and well-organized hyperlinked explanation of ecosystems, supported by a diagram showing the components of an ecosystem and their interrelatedness. Part of a much larger site 'Fundamentals of Physical Geography' set up by a Canadian University, the page covers many aspects of the area of study.

Landforms of Weathering
http://www.geog.ouc.bc.ca/physgeog/contents/11c.html

From a much larger site 'Fundamentals of Physical Geography' set up by a Canadian University and covering many aspects of the area of study. The site is clear and well organized and contains maps and diagrams.

Late Pleistocene Extinctions
http://www.museum.state.il.us/exhibits/larson/LP_extinction.html

Exploration of possible causes of the Late Pleistocene extinction of most large mammals in North America. Different theories are discussed. There is also information on the prehistoric inhabitants of North America.

Mapquest
http://www.mapquest.com/

Features interactive maps of more than 3 million cities worldwide. You can track down businesses and landmarks in Europe and zoom in on street maps. If you've lost your way, 'TripQuest' provides door-to-door driving directions.

Map Quiz Tutorial: Physical Geography
http://www.harper.cc.il.us/mhealy/mapquiz/intro/inphyfr.htm

Interactive site that invites you to locate specific features on a physical map of the world. The answer frame at the bottom of the page tells you if you were correct.

Mariana Trench, Pacific Ocean
http://automap.msn.com/am/pages/a/20293.htm

Animated journey through the deepest ocean trench in the world, in a series of downloadable MPEG animations. The animations are based on NASA data and this site also includes an illustrated geology of the area and a description of plate tectonics.

Mathematics of Cartography
http://math.rice.edu/~lanius/pres/map/

History of maps and details of the mathematics behind mapmaking. There are maths-related mapping problems to solve and a list of cartographical links.

Met. Office Home Page
http://www.meto.govt.uk/

Authoritative account of global warming issues such as the ozone problem, El Niño (and the less known La Niña) the tropical cyclones, and forecasting methods. Scientific explanations alternate with images and film clips in an educational site which especially targets teachers and their students.

Mineralogy Database
http://webmineral.com

Created by a mineral enthusiast, this is an extremely useful Web site if you need to know about a particular mineral. All major mineralogical data of each of the more than 3,700 minerals listed in *Dana's Mineralogy*, the definitive source on minerals, is in this searchable database.

Models of Landform Development
http://www.geog.ouc.bc.ca/physgeog/contents/11a.html

This page on landforms discusses the four types to be found on Earth, and how they develop. Part of a much larger site, 'Fundamentals of Physical Geography', set up by a Canadian University, the text is supported by a graphical model explaining the relationship between geomorphic processes and landform types.

Modified Mercalli Intensity Scale
http://wwwneic.cr.usgs.gov/neis/general/handouts/mercalli.html

Explanation of the Mercalli scale, which is one of the two main scales (with the Richter scale) along which earthquakes are measured.

MTU Volcanoes Page
http://www.geo.mtu.edu/volcanoes/

Provided by Michigan Technological University, this site includes a world map of volcanic activity with information on recent eruptions, the latest research in remote sensing of volcanoes, and many spectacular photographs.

Multimedia History of Glacier Bay, Alaska
http://sdcd.gsfc.nasa.gov/GLACIER.BAY/glacierbay.html

Virtual tour featuring scenic flights over 3-dimensional glaciers. The multimedia experience combines video footage, computer and satellite images, photos, text, and maps, and includes information on glacial formation. View hand-drawn maps of glaciers that date back to 1794 and watch video of massive ice fronts splitting and splashing into the sea.

Naming of Atlantic Hurricanes
http://www.nws.noaa.gov/er/box/hurricane-names.html

Interesting explanation about how hurricanes are named, from the US national weather service. There is also a table that shows you what name future hurricanes will be given.

National Geographic Online
http://www.nationalgeographic.com

Large and lavishly illustrated Web site. Features include the *Map Machine Atlas*, which allows you to find maps, flags, facts, and profiles for the countries of the world, and discussion forums on a variety of subjects. Many articles and multimedia items can be accessed for free.

Nature Explorer
http://www.naturegrid.org.uk/explorer/index.html

Wealth of information about water creatures. The presentation is fun and interactive, including a 'Virtual pond dip'.

NOAA
http://www.noaa.gov/

The US government National Oceanic and Atmospheric Administration (NOAA) site is beautiful and user-friendly, with information and the latest news on weather, climate, oceanography, and remote sensing.

OceanLink
http://oceanlink.island.net/index.html

>Good site for finding out about many marine animals, with encyclopedic descriptions of selected animals. This site contains sections called 'Kidlinks', 'Ask a scientist', 'Spotlight on...', and 'Ocean news'.

Ocean Planet
http://seawifs.gsfc.nasa.gov/ocean_planet.html

>Oceans and the environmental issues affecting their health, based on the Smithsonian Institute's travelling exhibition of the same name. Use the exhibition floor plan to navigate your way around the different 'rooms' – with themes ranging from ocean science and immersion to heroes and sea people.

Ocean Satellite Image Comparison
http://www.csc.noaa.gov/crs/real_time/javaprod/satcompare.html

>Maintained by the US National Oceanic Atmospheric Administration, this site is a Java applet which compares satellite images and data about the ocean surface temperatures and turbidity of numerous coastal areas around the USA. The site includes an image panner for navigating larger images.

Ordnance Survey
http://www.ordsvy.gov.uk/home/index.html

>Impressive site from the official body responsible for the mapping of Britain. Here you can download a selection of UK maps in a variety of formats. Other features include the 'Education' section which has a selection of teaching resources linked to maps and geography.

Out of This World Exhibition
http://www.lhl.lib.mo.us/pubserv/hos/stars/welcome.htm

>Educational and entertaining exhibition of celestial atlases – highly illustrated scientific books of the post-Renaissance period designed in an effort to capture 'the sweeping grandeur of the heavens, super-imposed with constellation figures, in a grand and monumental format'. As well as plenty of images, the site also offers an introductory historical essay and accompanying notes for each image.

PALEOMAP project
http://www.scotese.com/

>The goal of the PALEOMAP Project is to illustrate the plate tectonic development of the ocean basins and continents, as well as the changing distribution of land and sea during the past 1,100 million years.

Plate Tectonics
http://www.seismo.unr.edu/ftp/pub/louie/class/100/plate-tectonics.html

Well-illustrated site on this geological phenomenon. As well as the plentiful illustrations, this site also has a good clear manner of explaining the way the plates of the Earth's crust interact to produce seismic activity.

Plate Tectonics
http://volcano.und.nodak.edu/vwdocs/vwlessons/plate_tectonics/introduction.html

US-based site aimed at young people as a 'lesson covering the chemical and physical layers of the Earth, historical development of the theory, and descriptions of the location and types of plate boundaries.'

Primary Geography Page
http://users.netmatters.co.uk/tpickford/gg/tony.html

A site of links to pages providing resources, including the author's own software, for teachers and pupils of Key Stages 1 and 2 geography.

Questions and Answers About Snow
http://www-nsidc.colorado.edu/NSIDC/EDUCATION/SNOW/snow_FAQ.html

Comprehensive information about snow from the US National Snow and Ice Data Centre. Among the interesting subjects discussed are why snow is white, why snow flakes can be up to two inches across, what makes some snow fluffy, why sound travels farther across snowy ground, and why snow is a good insulator.

Rader's Terrarum
http://www.kapili.com/terrarum/index.html

The Terrarum bills itself as a resource on physical geography, but its emphasis on the relationships between the various forces such as energy, tectonics and climate brings in information on many areas of earth science. It would be a useful source for GCSE geographers, although the child-friendly language gives it the appearance of being designed for younger students.

Rainforests of the World
http://www.eco-portal.com/

Resource that explores many aspects of this type of vegetation. The site includes detail on 'Climate', 'Homeland for forest peoples', and 'Rainforest facts'.

River Gauging Stations
http://www.environment-agency.gov.uk/gui/dataset4/4frame.htm

UK Environment Agency page providing flow statistics for 35 English and Welsh rivers. Click on a region on the map of England and Wales, then select a gauging station to access flow statistics for individual rivers, as well as data on the geology of the catchment area and average monthly flow figures.

Rivers and Streams
http://www.wsu.edu/~geology/geol101/rivers/rivers.htm

Essential terms and concepts for the study of rivers, part of a university geography site. Hypertext is used to link definitions of drainage patterns, oxbow lakes, meanders, and other aspects of river formation, to simple sketch diagrams.

Royal Geographical Society
http://www.rgs.org/

Mine of information for both geographers and non-specialists, including events organized by the Royal Geographical Society, online exhibitions, field expeditions and research projects, publications, and links to other geographical organizations.

San Andreas Fault and Bay Area
http://sepwww.stanford.edu/oldsep/joe/fault_images/BayAreaSanAndreasFault.html

Detailed tour of the San Andreas Fault and the San Francisco Bay area, with information on the origination of the fault. The site is supported by a full range of area maps.

Science Odyssey: You Try It: Plate Tectonics
http://www.pbs.org/wgbh/aso/tryit/tectonics/intro.html

Well-presented site to 'the theory that Earth's outer layer is made up of plates'. This site, labelled 'Mountain maker, Earth shaker' has science-related interactive activities that use the Shockwave plug-in.

Soil pH – What It Means
http://www.esf.edu/pubprog/brochure/soilph/soilph.htm

Explanation of soil pH. The Web site also describes how to measure the pH of soil using simple experimental equipment, and goes on to describe methods that may be used to modify the acidity of alkalinity of your soil.

Solar Energy: Basic Facts
http://www.brookes.ac.uk/other/uk-ises/facts.htm

Quick guide to this renewable energy source. Taken from a series of fact sheets published by the Solar Energy Society, this page explains concepts such as solar water heating, photovoltaics, and passive solar architecture. Find out why solar power works better in Scotland than the South of France.

Some Like it Hot
http://www.blm.gov/education/sonoran/sonoran.html

Sponsored by the US Bureau of Land Management, this site explores the Sonoran Desert, the USA's hottest desert, and the unusual plant and animal life that has adapted to this harsh environment. Articles investigate management challenges facing the area, threatened and endangered species, and Native American cultural areas of the Sonoran Desert.

Space.Com
http://www.space.com/

Latest news on space exploration plus space quizzes, games, and other fun stuff. The site also includes a children's site with activities.

State of the Climate: a Time for Action
http://www.panda.org/climate_event/intro.htm

Climate change is happening now and having a drastic effect upon the natural world, according to this dramatic, animated Web site from the Worldwide Fund for Nature (WWF). A Macromedia Flash multimedia introduction is followed by information on WWF's climate change campaign.

Storm Chaser Home Page
http://www.geocities.com/CapeCanaveral/8546/nc.htm

Set up by a devotee of the sport, this site has swiftly expanded through voluntary contributions of other fans. There are a vast number of stunning photos and a whole online Tornado Museum to tempt you into this hobby. On the other hand, an extended section of veterans' thoughts on safety may make you think twice. The site also offers severe weather reports for emergencies!

Streamflow and Fluvial Processes
http://www.geog.ouc.bc.ca/physgeog/contents/11i.html

Clear, hyperlinked explanation of streamflow, which is part of a university site on physical geography. The site uses diagrams and photographs to illustrate the various fluvial processes.

Temporal Urban Mapping
http://edcwww2.cr.usgs.gov/umap/htmls/backgrnd.html

> Part of the US Geological Survey 'Urban Dynamics Research Program', this site covers the problems associated with 'Urban sprawl', 'Rates of change', and 'Environmental impacts'.

The Greenhouse Effect: How the Earth Stays Warm
http://www.enviroweb.org/edf/ishappening/greeneffect/index.html

> Explanation of the greenhouse effect, the process by which atmospheric gases trap heat. This page links to a description of how this perfectly normal and essential process is being turned into something harmful by the pollutants humans put into the atmosphere. It also explains that related phenomenon, the hole in the ozone layer.

The Watershed Game
http://www1.umn.edu/bellmuse/mnideals/watershed/watershed2.html

> Interactive game designed to teach pupils from secondary level upwards how human activity can affect the quality of the local water supply. Click on a section: national parks, agriculture, cities, or neighbourhoods, to access virtual reality simulations allowing you to make your own land-use decisions and see their impact on the watershed.

This Dynamic Earth: The Story of Plate Tectonics
http://pubs.usgs.gov/publications/text/dynamic.html

> Electronic version of a book published by the US Geological Survey.

This Dynamic Planet
http://pubs.usgs.gov/pdf/planet.html

> Online publication about plate tectonics, its history and theory, with a map that shows the Earth's physiographic features, the current movements of its major tectonic plates, and the locations of its volcanoes, earthquakes, and impact craters. The use of colour and shaded relief helps the reader to identify significant features of the land surface and the ocean floor. Over 1,500 volcanoes active during the past 10,000 years are plotted on the map in four age categories.

Thunderstorms and Tornadoes
http://www.geog.ouc.bc.ca/physgeog/contents/7t.html

> Striking photographs, one of a 21-cm hailstone, another of the anvil-shaped cloud typical of a severe thunderstorm, illustrate this hyperlinked description of some dramatic weather events. Part of a much larger physical geography site set up by a Canadian university, the page gives a clear explanation of the various types of thunderstorm and the conditions that bring them about.

Tornado Project Online
http://www.tornadoproject.com/

All about tornadoes – including myths and oddities, personal experiences, tornado chasing, tornado safety, and tornadoes past and present.

Tropical Savannas CRC: Landscape Processes
http://savanna.ntu.edu.au/information/information.html

Australian-based site providing information on tropical savannahs including 'Savannah explorer', 'Savannah map maker', where you can explore relationships between landscape features such as vegetation and soils by overlaying maps, and 'Savannah search'.

Tropical Weather and Hurricanes
http://www.geog.ouc.bc.ca/physgeog/contents/7u.html

From a much larger site 'Fundamentals of physical geography', set up by a Canadian university and covering many aspects of the area of study. The site includes details about how and where tropical storms develop. It is well organized and contains maps and diagrams.

Tsunami!
http://www.geophys.washington.edu/tsunami/intro.html

Description of many aspects of tsunamis. Included are details on how a tsunami is generated and how it propagates, how they have affected humans, and how people in coastal areas are warned about them. The site also discusses if and how you may protect yourself from a tsunami and provides 'near real-time' tsunami information bulletins.

Types of Rocks: Igneous, Metamorphic and Sedimentary
http://www.zephryus.demon.co.uk/virtualschool/lessons/Lesson008.htm

Targeted specifically at GCSE geographers, this clear introduction to the three main rock types begins by outlining the history of the rock cycle and the relationship between igneous, metamorphic, and sedimentary rocks. Clear explanations of all three types, with photographs to illustrate, are followed by a set of test problems to make you do your own research.

UK and Ireland Climate Index
http://www.onlineweather.com/BritishIsles/climate.html

Highlights of this very comprehensive climate information site include an index of the main UK and Irish weather stations, with tables of monthly minimum and maximum temperatures, as well as rainfall figures. This is also where to come to find the likelihood of rain in a host of towns from Barnstaple to Ullapool.

Unexplored Spaces
http://magma.nationalgeographic.com/2000/biodiversity/biomes/index.cfm

Visually stunning *National Geographic* page exploring some of the world's little-known biomes, and showing how much they have to teach scientists about the way in which life is organized. Taking a cobble-bed off the coast of New England, USA, as its starting point, this wide-ranging article branches off into the ocean floor, outer space and other unsuspected sources of knowledge about the processes of life.

Urban-Rural Population Distribution
http://www.csudh.edu/global_options/375Students-Sp96/Spain/URB.RURALPOP.HTML

General information on the urban–rural population distribution in the agricultural, industrial, and information ages, with extra insight on Spain.

US Geological Survey
http://www.usgs.gov/

US Geological Survey (USGS) site includes information on almost every aspect of earth science you can think of. It is well-organized and contains a useful search engine to make finding things very easy.

USGS Earthshots: Satellite Images of Environmental Change
http://edcwww.cr.usgs.gov/earthshots/slow/tableofcontents

See how Earth's environment is changing by accessing this fascinating collection of satellite pictures from the US Geological Survey (USGS). Start by clicking on a highlighted area on a world map to go to a series of detailed comparative images, or select from a huge collection of articles on subjects such as geology, water, and wildlife.

U-Shaped Valleys and Truncated Spurs
http://www.zephryus.demon.co.uk/geography/resources/glaciers/ushape.html

Information page aimed at GCSE-level geography students, which describes these important physical features and explains the process of glacial erosion which brought them into being.

Vesuvius, Italy
http://volcano.und.nodak.edu/vwdocs/volc_images/img_vesuvius.html

Site examining the complex geology of Vesuvius and its famous eruption of 79 AD. There are images of the volcano and historical drawings. There is also a link to a local site campaigning for an improved civil defence plan as the volcano prepares once more to explode.

Web Sites

View of a Sustainable World
http://www.environment.gov.au/education/aeen/pd/tsw/module1/module1.htm

Part of an electronic teacher's guide 'Teaching for a Sustainable World: International Edition', this site also contains useful information for individual students.

Virtual Cave
http://library.advanced.org/2974/

Browse the mineral wonders unique to the cave environment – from bell canopies and bottlebrushes to splattermites and stalactites.

Virtual Fieldwork
http://www.users.globalnet.co.uk/~drayner/indexvf.htm

Field studies are the best way to learn about geographical concepts; take the next best thing – an Internet field trip – by selecting from this page. The choices include coastal erosion in Kent, UK, and coastal sand dunes and management in North Wales, UK. Study digital photographs of the area, read the accompanying notes, then take a short test to find out how much you have learned.

Volcanoes
http://www.learner.org/exhibits/volcanoes/

Dramatic site that opens with a picture of an erupting volcano, and asks whether volcanic catastrophes – which threaten millions of people in an increasingly densely-populated world – can be predicted. With the help of video clips and interactive tasks, the well-written text goes on to answer important questions about volcanic activity: how volcanoes form and what the forces are that cause solid rock to melt and burst through the surface of the Earth.

VolcanoWorld
http://volcano.und.edu

Comprehensive site on volcanoes, with details of the most recent eruptions, currently active volcanoes, a glossary, images and video clips, and even a list of extraterrestrial volcanoes. If you can't find out what you want to know, you can 'Ask a volcanologist'.

Weather: What forces affect our weather?
http://www.learner.org/exhibits/weather/

User-friendly, US-based site exploring aspects of climate including 'the water cycle', 'powerful storms', and 'our changing climate'. The site includes 'hands-on' activities such as 'Try your hand at tornado chasing'.

Welcome to Coral Forest
http://www.blacktop.com/coralforest/

> Site dedicated to explaining the importance of coral reefs for the survival of the planet. It is an impassioned plea on behalf of the world's endangered coral reefs and includes a full description of their biodiversity, maps of where coral reefs are to be found (no less than 109 countries), and many photos.

Woods Hole Oceanographic Institution Home Page
http://www.whoi.edu/index.html

> Site run by a Massachusetts-based oceanographic institute. As well as containing details of their research programmes and an overview of the organization, there is a gallery of marine pictures and videos, and contacts for their education programmes.

World Environmental Changes Landmarks
http://www.bbc.co.uk/education/landmarks/

> Well-designed and attractive page from the BBC's education site looking at the effect the increasing human population has on the environment and climate. The site is broken down into five different environment types (wetlands, drylands, cities, the sea, and forests) with information, case studies from around the world, and activities for each. The high-tech version also includes animations.

World Meteorological Organization
http://www.wmo.ch/

> Internet voice of the World Meteorological Organization, a UN division coordinating global scientific activity related to climate and weather. The site offers ample material on the long-term objectives and immediate policies of the organization. It also disseminates important information on WMO's databases, training programmes, and satellite activities, as well as its projects related to the protection of the environment.

World of Amber
http://www.emporia.edu/S/www/earthsci/amber/amber.htm

> Everything you need to know about amber is here. There is information on its physical properties, uses, and geological and geographical occurrences, plus fossils in amber, recovery methods, amber myths, museums, and a quiz.

Worldtime
http://www.worldtime.com/

Interactive world atlas featuring information on local time and sunrise and sunsets in hundreds of cities, and a database of public holidays around the world. You can rotate the globe to view areas of daylight and night, zoom in on areas of interest, or display national borders.

Worldwide Earthquake Locator
http://www.geo.ed.ac.uk/quakes/quakes.html

Edinburgh University, Scotland, runs this site, which allows visitors to search for the world's latest earthquakes. The locator works on a global map on which perspective can be zoomed in or out. There are normally around five or six earthquakes a day; you'll find it surprising how few make the news. The site also has some general information on earthquakes.

7 Glossary

ablation
loss of snow and ice from a glacier by melting and evaporation. It is the opposite of accumulation.
   Ablation is most significant near the snout, or foot, of a glacier, since temperatures tend to be higher at lower altitudes. The rate of ablation also varies according to the time of year, being greatest during the summer. If total ablation exceeds total accumulation for a particular glacier, then the glacier will retreat, and vice versa.

abrasion
effect of corrasion, a type of erosion in which rock fragments scrape and grind away a surface. The rock fragments may be carried by rivers, wind, ice, or the sea. Striations, or grooves, on rock surfaces are common abrasions, caused by the scratching of rock debris embedded in glacier ice.

abrasive
substance used for cutting and polishing or for removing small amounts of the surface of hard materials. There are two types: natural and artificial abrasives, and their hardness is measured using the Mohs scale. Natural abrasives include quartz, sandstone, pumice, diamond, emery, and corundum; artificial abrasives include rouge, whiting, and carborundum.

abyssal plain
broad, relatively flat expanse of sea floor lying 3–6 km/2–4 mi below sea level. Abyssal plains are found in all the major oceans, and they extend from bordering continental rises to mid-oceanic ridges. Abyssal plains are covered in a thick layer of sediment, and their flatness is punctuated by rugged low abyssal hills and high sea mounts.

abyssal zone
dark ocean region 2,000–6,000 m/6,500–19,500 ft deep; temperature 4°C/39°F. Three-quarters of the area of the deep-ocean floor lies in the abyssal zone, which is too far from the surface for photosynthesis to take place. Some fish and crustaceans living there are blind or have their own light sources. The region above is the bathyal zone; the region below, the hadal zone.

acid rock
igneous rock that contains more than 60% by weight silicon dioxide, SiO_2, such as a granite or rhyolite. Along with the terms *basic rock* and *ultrabasic rock* it is part of an outdated classification system based on the erroneous belief that

silicon in rocks is in the form of silicic acid. Geologists today are more likely to use the descriptive term felsic rock or report the amount of SiO_2 in percentage of weight.

aclinic line
magnetic equator, an imaginary line near the Equator, where a compass needle balances horizontally, the attraction of the north and south magnetic poles being equal.

advection fog
fog formed by warm moist air meeting a cold ocean current or flowing over a cold land surface. It is common in coastal areas where warm air from the sea meets the colder land surface, particularly at night or during the winter.

agate
cryptocrystalline (with crystals too small to be seen with an optical microscope) silica, SiO_2, composed of cloudy and banded chalcedony, sometimes mixed with opal, that forms in rock cavities.

air mass
large body of air with particular characteristics of temperature and humidity. An air mass forms when air rests over an area long enough to pick up the conditions of that area. When an air mass moves to another area it affects the weather of that area, but its own characteristics become modified in the process. For example, an air mass formed over the Sahara will be hot and dry, becoming cooler as it moves northwards. Air masses that meet form *fronts*.

There are four types of air masses. *Tropical continental* (Tc) air masses form over warm land and *Tropical maritime* (Tm) masses form over warm seas. Air masses that form over cold land are called *Polar continental* (Pc) and those forming over cold seas are called *Polar* or *Arctic maritime* (Pm or Am).

The weather of the UK is affected by a number of air masses which, having different characteristics, bring different weather conditions. For example, an Arctic air mass brings cold conditions whereas a Saharan air mass brings hot conditions.

alluvial fan
roughly triangular sedimentary formation found at the base of slopes. An alluvial fan results when a sediment-laden stream or river rapidly deposits its load of gravel and silt as its speed is reduced on entering a plain.

The surface of such a fan slopes outward in a wide arc from an apex at the mouth of the steep valley. A small stream carrying a load of coarse particles builds a shorter, steeper fan than a large stream carrying a load of fine particles. Over time, the fan tends to become destroyed piecemeal by the continuing headward and downward erosion levelling the slope.

alluvium
fine silty material deposited by a river. It is deposited along the river channel where the water's velocity is too low to transport the river's load – for example, on the inside bend of a meander. A flood plain is composed of alluvium periodically deposited by floodwater. Sometimes alluvial deposits after flooding result in oxbow lakes. A deposit at the mouth of a river entering the sea is known as marine alluvium.

altimetry
process of measuring altitude, or elevation. Satellite altimetry involves using an instrument – commonly a laser – to measure the distance between the satellite and the ground.

amethyst
variety of quartz, SiO_2, coloured violet by the presence of small quantities of impurities such as manganese or iron; used as a semiprecious stone. Amethysts are found chiefly in the Ural Mountains, India, the USA, Uruguay, and Brazil.

anabatic wind
warm wind that blows uphill in steep-sided valleys in the early morning. As the sides of a valley warm up in the morning the air above is also warmed and rises up the valley to give a gentle breeze. By contrast, a katabatic wind is cool and blows down a valley at night.

aneroid barometer
see barometer.

anthracite
hard, dense, shiny variety of coal, containing over 90% carbon and a low percentage of ash and impurities, which causes it to burn without flame, smoke, or smell. Because of its purity, anthracite gives off relatively little sulphur dioxide when burnt.

anticyclone
area of high atmospheric pressure caused by descending air, which becomes warm and dry. Winds radiate from a calm centre, taking a clockwise direction in the northern hemisphere and an anticlockwise direction in the southern hemisphere. Anticyclones are characterized by clear weather and the absence of rain and violent winds. In summer they bring hot, sunny days and in winter they bring fine, frosty spells, although fog and low cloud are not uncommon in the UK.

Blocking anticyclones, which prevent the normal air circulation of an area, can cause summer droughts and severe winters. For example, the summer drought in Britain 1976, and the severe winters of 1947 and 1963 were caused by blocking anticyclones.

Glossary

aquamarine
blue variety of the mineral beryl. A semiprecious gemstone, it is used in jewellery.

aquifer
body of rock through which appreciable amounts of water can flow. The rock of an aquifer must be porous and permeable (full of interconnected holes) so that it can conduct water. Aquifers are an important source of fresh water, for example, for drinking and irrigation, in many arid areas of the world, and are exploited by the use of artesian wells.

An aquifer may be underlain, overlain, or sandwiched between less permeable layers, called aquicludes or *aquitards*, which impede water movement. Sandstones and porous limestones make the best aquifers.

Archaean or Archaeozoic
widely used term for the earliest era of geological time; the first part of the Precambrian *Eon*, spanning the interval from the formation of Earth to about 2,500 million years ago.

arête German grat; North American combe-ridge
sharp narrow ridge separating two glacial troughs (valleys), or corries. The typical U-shaped cross sections of glacial troughs give arêtes very steep sides. Arêtes are common in glaciated mountain regions such as the Rockies, the Himalayas, and the Alps.

artesian well
well that is supplied with water rising naturally from an underground water-saturated rock layer (aquifer). The water rises from the aquifer under its own pressure. Such a well may be drilled into an aquifer that is confined by impermeable rocks both above and below. If the water table (the top of the region of water saturation) in that aquifer is above the level of the well head, hydrostatic pressure will force the water to the surface.

Artesian wells are often overexploited because their water is fresh and easily available, and they eventually become unreliable. There is also some concern that pollutants such as pesticides or nitrates can seep into the aquifers.

Much use is made of artesian wells in eastern Australia, where aquifers filled by water in the Great Dividing Range run beneath the arid surface of the Simpson Desert. The artesian well is named after Artois, a French province, where the phenomenon was first observed.

asbestos
any of several related minerals of fibrous structure that offer great heat resistance because of their nonflammability and poor conductivity. Commercial asbestos is generally either made from serpentine ('white' asbestos) or from sodium iron silicate ('blue' asbestos). The fibres are woven together or bound by an inert material. Over time the fibres can work loose and, because they are

small enough to float freely in the air or be inhaled, asbestos usage is now strictly controlled; exposure to its dust can cause cancer.

aspect
direction in which a slope faces. In the northern hemisphere a slope with a southerly aspect receives more sunshine than other slopes and is therefore better suited for growing crops that require many hours of sunshine in order to ripen successfully. Vineyards in northern Europe are usually situated on south-facing slopes.

asthenosphere
layer within Earth's mantle lying beneath the lithosphere, typically beginning at a depth of approximately 100 km/63 mi and extending to depths of approximately 260 km/160 mi. Sometimes referred to as the 'weak sphere', it is characterized by being weaker and more elastic than the surrounding mantle.

The asthenosphere's elastic behaviour and low viscosity allow the overlying, more rigid plates of lithosphere to move laterally in a process known as plate tectonics. Its elasticity and viscosity also allow overlying crust and mantle to move vertically in response to gravity to achieve *isostatic equilibrium*.

atmosphere
mixture of gases surrounding a planet. Planetary atmospheres are prevented from escaping by the pull of gravity. Atmospheric pressure, the density of gases in the atmosphere, decreases with altitude. In its lowest layer, the Earth's atmosphere consists of nitrogen (78%) and oxygen (21%), both in molecular form (two atoms bonded together) and 1% argon. Small quantities of other gases are important to the chemistry and physics of the Earth's atmosphere, including water, carbon dioxide and ozone. The atmosphere plays a major part in the various cycles of nature (the water cycle, the carbon cycle, and the nitrogen cycle). It is the principal industrial source of nitrogen, oxygen, and argon, which are obtained by fractional distillation of liquid air.

The Earth's atmosphere is divided into four regions of atmosphere classified by temperature.

Troposphere
This is the lowest level of the atmosphere (altitudes from 0 to 10 km/6 mi) and it is heated to an average temperature of 15°C/59°F by the Earth, which in turn is warmed by infrared and visible radiation from the Sun. Warm air cools as it rises in the troposphere and this rising of warm air causes rain and most other weather phenomena. The top of the troposphere is approximately –60°C/ –140°F.

Stratosphere
Temperature increases with altitude in this next layer (from 10 km/6 mi to 50 km/31 mi), from –60°C/–140°F to near 0°C/32°F.

Glossary

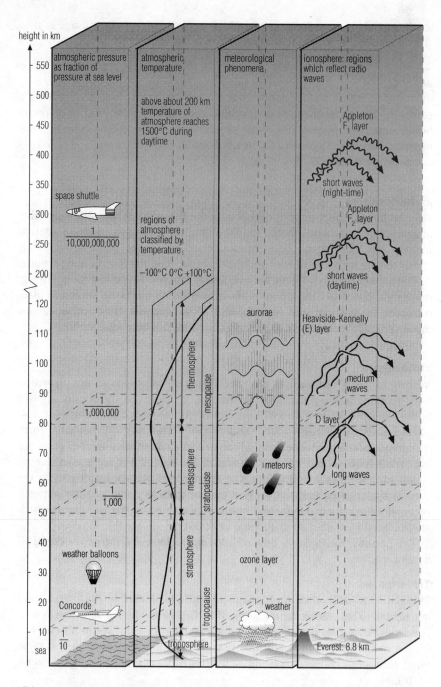

All but 1% of the Earth's atmosphere lies in a layer 30 km/19 mi above the ground. At a height of 5,500 m/18,000 ft, air pressure is half that at sea level. The temperature of the atmosphere varies greatly with height; this produces a series of layers, called the troposphere, stratosphere, mesosphere, and thermosphere.

Mesosphere
Temperature again decreases with altitude through the mesosphere (50 km/31 mi to 80 km/50 mi), from 0°C/32°F to below –100°C/–212°F.

Thermosphere
In the highest layer (80 km/50 mi to about 700 km/450 mi). Temperature rises with altitude to extreme values of thousands of degrees. The meaning of these extreme temperatures can be misleading. High thermosphere temperatures represent little heat because they are defined by motions among so few atoms and molecules spaced widely apart from one another.

atmospheric circulation
large-scale movement of air within the lower atmosphere. Warm air at the Equator rises, creating a zone of low pressure. This air moves towards the poles, losing energy and becoming cooler and denser, until it sinks back to the surface at around 30° latitude, creating an area of high pressure. At the surface, air moves from this high pressure zone back towards the low pressure zone at the equator, completing a circulatory movement.

atmospheric pressure
pressure at any point on the Earth's surface that is due to the weight of the column of air above it; it therefore decreases as altitude increases, simply because there is less air above. At sea level the average pressure is 101 kilopascals (1,013 millibars, 760 mmHg, or 14.7 lb per sq in, or 1 atmosphere). Changes in atmospheric pressure, measured with a barometer, are used in weather forecasting. Areas of relatively high pressure are called anticyclones; areas of low pressure are called depressions.

atoll
continuous or broken circle of coral reef and low coral islands surrounding a lagoon.

attrition
process by which particles of rock being transported by river, wind, or sea are rounded and gradually reduced in size by being struck against one another.

The rounding of particles is a good indication of how far they have been transported. This is particularly true for particles carried by rivers, which become more rounded as the distance downstream increases.

avalanche
fall or flow of a mass of snow and ice down a steep slope under the force of gravity. Avalanches occur because of the unstable nature of snow masses in mountain areas.

Changes of temperature, sudden sound, or earth-borne vibrations may trigger an avalanche, particularly on slopes of more than 35°. The snow compacts into ice as it moves, and rocks may be carried along, adding to the damage caused.

Avalanches leave slide tracks, long gouges down the mountainside that can be up to 1 km/0.6 mi long and 100 m/330 ft wide. These slides have a similar beneficial effect on biodiversity as do forest fires, clearing the land of snow and mature mountain forest, enabling plants and shrubs that cannot grow in shade, to recolonize; and creating wildlife corridors.

backwash
retreat of a wave that has broken on a beach. When a wave breaks, water rushes up the beach as swash and is then drawn back towards the sea as backwash.

bar
deposit of sand or silt formed in a river channel, or a long sandy ridge running parallel to a coastline that is submerged at high tide. Coastal bars can extend across estuaries to form *bay bars*. These bars are greatly affected by the beach cycle. The high tides and high waves of winter erode the beach and deposit the sand in bars offshore.

barometer
instrument that measures atmospheric pressure as an indication of weather. Most often used are the *mercury barometer* and the *aneroid barometer*.

In a mercury barometer a column of mercury in a glass tube, roughly 0.75 m/2.5 ft high (closed at one end, curved upwards at the other), is balanced by the pressure of the atmosphere on the open end; any change in the height of the column reflects a change in pressure. In an aneroid barometer, a shallow cylindrical metal box containing a partial vacuum expands or contracts in response to changes in pressure.

basalt
commonest volcanic igneous rock in the Solar System. Much of the surfaces of the terrestrial planets Mercury, Venus, Earth, and Mars, as well as the Moon, are composed of basalt. Earth's ocean floor is virtually entirely made of basalt. Basalt is mafic, that is, it contains relatively little silica: about 50% by weight. It is usually dark grey but can also be green, brown, or black. Its essential constituent minerals are calcium-rich feldspar and calcium and magnesium-rich pyroxene.

baseflow or groundwater flow
movement of water from land to river through rock. It is the slowest form of such water movement, and accounts for the constant flow of water in rivers during times of low rainfall, and makes up the river's base line on a hydrograph.

base level
level, or altitude, at which a river reaches the sea or a lake. The river erodes down to this level. If base level falls (due to uplift or a drop in sea level), rejuvenation takes place.

basic rock

igneous rock with relatively low silica contents of 45–52% by weight, such as gabbro and basalt. Along with the terms *acid rock* and *ultrabasic rock* it is part of an outdated classification system based on the erroneous belief that silicon in rocks is in the form of silicic acid. Geologists today are more likely to use the descriptive term mafic rock or report the amount of silicon dioxide (SiO_2) in percentage of weight.

batholith

large, irregular, deep-seated mass of intrusive igneous rock, usually granite, with an exposed surface of more than 100 sq km/40 sq mi. The mass forms by the intrusion or upswelling of magma (molten rock) through the surrounding rock. Batholiths form the core of some large mountain ranges like the Sierra Nevada of western North America.

According to plate tectonic theory, magma rises in subduction zones along continental margins where one plate sinks beneath another. The solidified magma becomes the central axis of a rising mountain range, resulting in the deformation (folding and overthrusting) of rocks on either side. Gravity measurements indicate that the downward extent or thickness of many batholiths is some 6–9 mi/10–15 km.

In the UK, a batholith underlies SW England and is exposed in places to form areas of high ground such as Dartmoor and Land's End.

bathyal zone

upper part of the ocean, which lies on the continental shelf at a depth of between 200 m/650 ft and 2,000 m/6,500 ft.

Bathyal zones (both temperate and tropical) have greater biodiversity than coral reefs, according to a 1995 study by the Natural History Museum in London. Maximum biodiversity occurs between 1,000 m/3,280 ft and 3,000 m/9,800 ft.

bauxite

principal ore of aluminium, consisting of a mixture of hydrated aluminium oxides and hydroxides, generally contaminated with compounds of iron, which give it a red colour. It is formed by the chemical weathering of rocks in tropical climates. Chief producers of bauxite are Australia, Guinea, Jamaica, Russia, Kazakhstan, Suriname, and Brazil.

Beaufort scale

system of recording wind velocity, devised by Francis Beaufort in 1806. It is a numerical scale ranging from 0 to 17, calm being indicated by 0 and a hurricane by 12; 13–17 indicate degrees of hurricane force.

In 1874 the scale received international recognition; it was modified in 1926. Measurements are made at 10 m/33 ft above ground level.

Glossary

bed
single sedimentary rock unit with a distinct set of physical characteristics or contained fossils, readily distinguishable from those of beds above and below. Well-defined partings called **bedding planes**.

Benguela current
cold ocean current in the South Atlantic Ocean, moving northwards along the west coast of southern Africa and merging with the south equatorial current at a latitude of 15° S. Its rich plankton supports large, commercially exploited fish populations.

Benioff zone
seismically active zone inclined from a deep-sea trench. Earthquakes along Benioff zones apparently define lithospheric plate that descends in to the mantle beneath another, overlying plate. The zone is named after Hugo Benioff, a US seismologist who first described this feature.

benthic
term describing the environment on the sea floor supporting bottom-dwelling plants (such as seaweeds) and animals (including corals, anemones, sponges, and shellfish).

bergschrund
deep crevasse that may be found at the head of a glacier. It is formed as a glacier pulls away from the headwall of the corrie, or hollow, in which it accumulated.

Beringia or Bering Land Bridge
former land bridge 1,600 km/1,000 mi wide between Asia and North America; it existed during the ice ages that occurred before 35,000 BC and during the period 24,000–9,000 BC. As the climate warmed and the ice sheets melted, Beringia flooded. It is now covered by the Bering Strait and Chukchi Sea.

berm
on a beach, a ridge of sand or pebbles running parallel to the water's edge, formed by the action of the waves on beach material. Sand and pebbles are deposited at the farthest extent of swash (advance of water) to form a berm. Berms can also be formed well up a beach following a storm, when they are known as *storm berms*.

bight
coastal indentation, crescent-shaped or gently curving, such as the Bight of Biafra in West Africa and the Great Australian Bight.

biogeochemistry
emerging branch of geochemistry involving the study of how chemical elements and their isotopes move between living organisms and geological materials. For example, the analysis of carbon in bone gives biogeochemists information on how the animal lived, its diet, and its environment.

biological weathering or organic weathering
form of weathering caused by the activities of living organisms – for example, the growth of roots or the burrowing of animals. Tree roots are probably the most significant agents of biological weathering as they are capable of prising apart rocks by growing into cracks and joints.

bolson
basin without an outlet, found in desert regions. Bolsons often contain temporary lakes, called playa lakes, and become filled with sediment from inflowing intermittent streams.

bore
surge of tidal water up an estuary or a river, caused by the funnelling of the rising tide by a narrowing river mouth. A very high tide, possibly fanned by wind, may build up when it is held back by a river current in the river mouth. The result is a broken wave, a metre or a few feet high, that rushes upstream.

Famous bores are found in the rivers Severn (England), Seine (France), Hooghly (India), and Chang Jiang (China), where bores of over 4 m/13 ft have been reported.

Bouguer anomaly
anomaly in the local gravitational force that is due to the density of rocks rather than local topography, elevation, or latitude. A positive anomaly, for instance, is generally indicative of denser and therefore more massive rocks at or below the surface. A negative anomaly indicates less massive materials. Calculations of bouguer anomalies are used for mineral prospecting and for understanding the structure beneath the Earth's surface. The Bouguer anomaly is named after its discoverer, the French mathematician Pierre Bouguer (1698–1758), who first observed it 1735.

butte
steep-sided, flat-topped hill, formed in horizontally layered sedimentary rocks, largely in arid areas. A large butte with a pronounced tablelike profile is a mesa.

Buttes and mesas are characteristic of semi-arid areas where remnants of resistant rock layers protect softer rock underneath, as in the plateau regions of Colorado, Utah, and Arizona, USA.

calcite
colourless, white, or light-coloured common rock-forming mineral, calcium carbonate, $CaCO_3$. It is the main constituent of limestone and marble and forms many types of invertebrate shell.

Glossary

Chimney Rock near Shiprock, New Mexico, USA – a notable example of the geological formation the butte. K G Preston-Mafham/Premaphotos Wildlife

caldera

a very large basin-shaped crater. Calderas are found at the tops of volcanoes, where the original peak has collapsed into an empty chamber beneath. The basin, many times larger than the original volcanic vent, may be flooded, producing a crater lake, or the flat floor may contain a number of small volcanic cones, produced by volcanic activity after the collapse.

Typical calderas are Kilauea, Hawaii; Crater Lake, Oregon, USA; and the summit of Olympus Mons, on Mars. Some calderas are wrongly referred to as craters, such as Ngorongoro, Tanzania.

California current

cold ocean current in the eastern Pacific Ocean flowing southwards down the West coast of North America. It is part of the North Pacific gyre (a vast, circular movement of ocean water).

calima

dust cloud in Europe, coming from the Sahara Desert, which sometimes causes heatwaves and eye irritation.

Cambrian period

period of geological time roughly 570–510 million years ago; the first period of the Palaeozoic Era. All invertebrate animal life appeared and marine algae were widespread. The **Cambrian Explosion** 530–520 million years ago saw the major radiaton in the fossil record of modern animal phyla; the earliest fossils with hard shells, such as trilobites, date from this period.

The name comes from Cambria, the medieval Latin name for Wales, where Cambrian rocks are typically exposed and were first described.

Canaries current
cold ocean current in the North Atlantic Ocean flowing southwest from Spain along the northwest coast of Africa. It meets the northern equatorial current at a latitude of 20° N.

canyon
deep, narrow valley or gorge running through mountains. Canyons are formed by stream down-cutting, usually in arid areas, where the rate of down-cutting is greater than the rate of weathering, and where the stream or river receives water from outside the area.

There are many canyons in the western USA and in Mexico, for example the Grand Canyon of the Colorado River in Arizona, the canyon in Yellowstone National Park, and the Black Canyon in Colorado.

carbonation
form of chemical weathering caused by rainwater that has absorbed carbon dioxide from the atmosphere and formed a weak carbonic acid. The slightly acidic rainwater is then capable of dissolving certain minerals in rocks. Limestone is particularly vulnerable to this form of weathering.

Carboniferous period
period of geological time roughly 362.5–290 million years ago, the fifth period of the Palaeozoic Era. In the USA it is divided into two periods: the Mississippian (lower) and the Pennsylvanian (upper).

Typical of the lower-Carboniferous rocks are shallow-water limestones, while upper-Carboniferous rocks have delta deposits with coal (hence the name). Amphibians were abundant, and reptiles evolved during this period.

cartography
art and practice of drawing maps, originally with pens and drawing boards, but now mostly with computer aided drafting programmes.

cave
roofed-over cavity in the Earth's crust usually produced by the action of underground water or by waves on a seacoast. Caves of the former type commonly occur in areas underlain by limestone, such as Kentucky and many Balkan regions, where the rocks are soluble in water. A *pothole* is a vertical hole in rock caused by water descending a crack; it is thus open to the sky.

Limestone caves
Most inland caves are found in karst regions, because limestone is soluble when exposed to groundwater. As the water makes its way along the main joints, fissures, and bedding planes, they are constantly enlarged into potential cave

passages, which ultimately join to form a complex network. Stalactites and stalagmites and columns form due to water rich in calcium carbonate dripping from the roof of the cave. The collapse of the roof of a cave produces features such as *natural arches* and *steep-sided gorges*.

Limestone caves are usually found just below the water table, wherever limestone outcrops on the surface. The biggest cave in the world is over 70 km/43 mi long, at Holloch, Switzerland.

Sea caves

Coastal caves are formed where relatively soft rock or rock containing definite lines of weakness, like basalt at tide level, is exposed to severe wave action. The gouging process (corrasion) and dissolution (corrosion) of weaker, more soluble rock layers is exacerbated by subsidence, and the hollow in the cliff face grows still larger because of air compression in the chamber. Where the roof of a cave has fallen in, the vent up to the land surface is called a blowhole. If this grows, finally destroying the cave form, the outlying truncated 'portals' of the cave are known as stacks or columns. The Old Man of Hoy (137 m/449 ft high), in the Orkney Islands, is a fine example of a stack.

cement

cement refers to a chemically precipitated material such as carbonate that occupies the interstices of clastic rocks.

Cenozoic Era or Caenozoic

era of geological time that began 65 million years ago and continues to the present day. It is divided into the Tertiary and Quaternary periods. The Cenozoic marks the emergence of mammals as a dominant group, and the rearrangment of continental masses towards their present positions.

chalcedony

form of the mineral quartz, SiO_2, in which the crystals are so fine-grained that they are impossible to distinguish with a microscope (cryptocrystalline). Agate, onyx, and carnelian are gem varieties of chalcedony.

chalk

soft, fine-grained, whitish sedimentary rock composed of calcium carbonate, $CaCO_3$, extensively quarried for use in cement, lime, and mortar, and in the manufacture of cosmetics and toothpaste. **Blackboard chalk** in fact consists of gypsum (calcium sulphate, $CaSO_4 \cdot 2H_2O$).

chemical weathering

form of weathering brought about by chemical attack of rocks, usually in the presence of water. Chemical weathering involves the 'rotting', or breakdown, of the original minerals within a rock to produce new minerals (such as clay minerals). Some chemicals are dissolved and carried away from the weathering source, while others are brought in.

A number of processes bring about chemical weathering, including carbonation (breakdown by weakly acidic rainwater), hydrolysis (breakdown by water), hydration (breakdown by the absorption of water), and oxidation (breakdown by the oxygen in water). The reaction of carbon dioxide gas in the atmosphere with silicate minerals in rocks to produce carbonate minerals is called the 'Urey reaction' after the US chemist who proposed it, Harold Urey. The Urey reaction is an important link between Earth's climate and the geology of the planet. It has been proposed that chemical weathering of large mountain ranges like the Himalayas of Nepal can remove carbon dioxide from the atmosphere by the Urey reaction (or other more complicated reactions like it), leading to a cooler climate as the greenhouse effects of the lost carbon dioxide are diminished.

chinook

warm dry wind that blows downhill on the east side of the Rocky Mountains of North America. It often occurs in winter and spring when it produces a rapid thaw, and so is important to the agriculture of the area.

The chinook is similar to the föhn in the valleys of the European Alps.

cinnabar

mercuric sulphide mineral, HgS, the only commercially useful ore of mercury. It is deposited in veins and impregnations near recent volcanic rocks and hot springs. The mineral itself is used as a red pigment, commonly known as **vermilion**. Cinnabar is found in the USA (California), Spain (Almadén), Peru, Italy, and Slovenia.

cirque

French name for a corrie, a steep-sided hollow in a mountainside.

clay

very fine-grained sedimentary deposit that has undergone a greater or lesser degree of consolidation. When moistened it is plastic, and it hardens on heating, which renders it impermeable. It may be white, grey, red, yellow, blue, or black, depending on its composition. Clay minerals consist largely of hydrous silicates of aluminium and magnesium together with iron, potassium, sodium, and organic substances. The crystals of clay minerals have a layered structure, capable of holding water, and are responsible for its plastic properties. According to international classification, in mechanical analysis of soil, clay has a grain size of less than 0.002 mm/0.00008 in.

climate

combination of weather conditions at a particular place over a period of time – usually a minimum of 30 years. A classification of climate encompasses the averages, extremes, and frequencies of all meteorological elements such as temperature, atmospheric pressure, precipitation, wind, humidity, and sunshine, together with the factors that influence them. The primary factors involved are: latitude (as a result of the Earth's rotation and orbit); ocean currents; large-scale movements of wind belts and air masses over the Earth's surface; temperature differences between land and sea surfaces; topography; continent positions; and vegetation. In the long term, changes in the Earth's orbit and the angle of its axis inclination also affect climate.

Climatology, the scientific study of climate, includes the construction of computer-generated models, and considers not only present-day climates, their effects and their classification, but also long-term climate changes, covering both past climates (palaeoclimates) and future predictions. Climatologists are especially concerned with the influence of human activity on climate change, among the most important of which, at both a local and global level, are those currently linked with ozone depleters and the greenhouse effect.

Climate classification

The word climate comes from the Greek *klima*, meaning an inclination or slope (referring to the angle of the Sun's rays, and thus latitude) and the earliest known classification of climate was that of the ancient Greeks, who based their system on latitudes. In recent times, many different systems of classifying climate have been devised, most of which follow that formulated by the German climatologist Wladimir Köppen (1846–1940) in 1900. These systems use vegetation-based classifications such as desert, tundra, and rainforest. Classification by air mass is used in conjunction with this method. This idea was first introduced in 1928 by the Norwegian meteorologist Tor Bergeron, and links the climate of an area with the movement of the air masses it experiences.

In the 18th century, the British scientist George Hadley developed a model of the general circulation of atmosphere based on convection. He proposed a simple pattern of cells of warm air rising at the Equator and descending at the poles. In fact, due to the rotation of the Earth, there are three such cells in each hemisphere. The first two of these consist of air that rises at the Equator and sinks at latitudes north and south of the tropics; the second two exist at the mid-latitudes where the rising air from the sub-tropics flows towards the cold air masses of the third pair of cells circulating from the two polar regions. Thus, in this model, there are six main circulating cells of air above ground producing seven terrestrial zones. There are three rainy regions (at the Equator and the temperate latitudes) resulting from the moisture-laden rising air, interspersed and bounded by four dry or desert regions (at the poles and sub-tropics) resulting from the dry descending air.

climate change
change in the climate of an area or of the whole world over an appreciable period of time. That is, a single winter that is colder than average does not indicate climate change; it is the change in average weather conditions from one period of time (30–50 years) to the next.

climate model
computer simulation, based on physical and mathematical data, of a climate system, usually the global (rather than local) climate. It is used by researchers

to study such topics as the possible long-term disruptive effects of the greenhouse gases, or of variations in the amount of radiation given off by the Sun.

climatology

study of climate, its global variations and causes.

Climatologists record mean daily, monthly, and annual temperatures and monthly and annual rainfall totals, as well as maximum and minimum values. Other data collected relate to pressure, humidity, sunshine, cold cover, and the frequency of days of frost, snow, hail, thunderstorms, and gales. The main facts are summarized in tables and climatological atlases published by nearly all the national meteorological services of the world. Climatologists also study climates of the past (palaeoclimates) by gathering information from such things as tree rings, deep sea sediments, and ice cores – all of which record various climate factors as they form.

clint

one of a number of flat-topped limestone blocks that make up a limestone pavement. Clints are separated from each other by enlarged joints called grykes.

cloud

water vapour condensed into minute water particles that float in masses in the atmosphere. Clouds, like fogs or mists, which occur at lower levels, are formed by the cooling of air containing water vapour, which generally condenses around tiny dust particles.

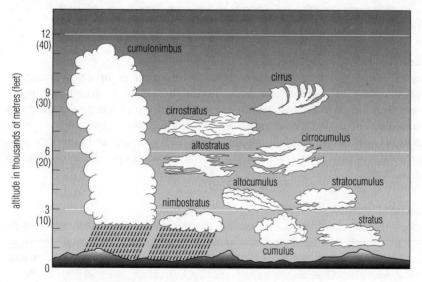

Standard types of cloud. The height and nature of a cloud can be deduced from its name. Cirrus clouds are at high levels and have a wispy appearance. Stratus clouds form at low level and are layered. Middle-level clouds have names beginning with 'alto'. Cumulus clouds, ball or cottonwool clouds, occur over a range of heights.

Glossary

Clouds are classified according to the height at which they occur and their shape. *Cirrus* and *cirrostratus* clouds occur at around 10 km/33,000 ft. The former, sometimes called mares' tails, consist of minute specks of ice and appear as feathery white wisps, while cirrostratus clouds stretch across the sky as a thin white sheet. Three types of cloud are found at 3–7 km/10,000–23,000 ft: cirrocumulus, altocumulus, and altostratus. *Cirrocumulus* clouds occur in small or large rounded tufts, sometimes arranged in the pattern called mackerel sky.

Altocumulus clouds are similar, but larger, white clouds, also arranged in lines. *Altostratus* clouds are like heavy cirrostratus clouds and may stretch across the sky as a grey sheet. *Stratocumulus* clouds are generally lower, occurring at 2–6 km/6,500–20,000 ft. They are dull grey clouds that give rise to a leaden sky that may not yield rain. Two types of clouds, *cumulus* and *cumulonimbus*, are placed in a special category because they are produced by daily ascending air currents, which take moisture into the cooler regions of the atmosphere. Cumulus clouds have a flat base generally at 1.4 km/4,500 ft where condensation begins, while the upper part is dome-shaped and extends to about 1.8 km/6,000 ft. Cumulonimbus clouds have their base at much the same level, but extend much higher, often up to over 6 km/20,000 ft. Short heavy showers and sometimes thunder may accompany them. *Stratus* clouds, occurring below 1–2.5 km/3,000–8,000 ft, have the appearance of sheets parallel to the horizon and are like high fogs.

In addition to their essential role in the water cycle, clouds are important in the regulation of radiation in the Earth's atmosphere. They reflect short-wave radiation from the Sun, and absorb and re-emit long-wave radiation from the Earth's surface.

coal

black or blackish mineral substance formed from the compaction of ancient plant matter in tropical swamp conditions. It is used as a fuel and in the chemical industry. Coal is classified according to the proportion of carbon it contains. The main types are **anthracite** (shiny, with about 90% carbon), **bituminous coal** (shiny and dull patches, about 75% carbon), and **lignite** (woody, grading into peat, about 50% carbon). Coal burning is one of the main causes of acid rain.

coastal erosion

erosion of the land by the constant battering of the sea's waves, primarily by the processes of hydraulic action, corrasion, attrition, and corrosion. Hydraulic action occurs when the force of the waves compresses air pockets in coastal rocks and cliffs. The air expands explosively, breaking the rocks apart. Rocks and pebbles flung by waves against the cliff face wear it away by the process of corrasion. Chalk and limestone coasts are often broken down by solution (also called corrosion). Attrition is the process by which the eroded rock particles themselves are worn down, becoming smaller and more rounded.

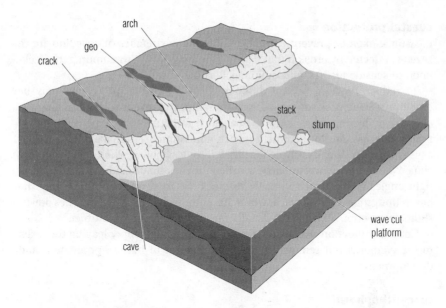

Typical features of coastal erosion: from the initial cracks in less resistant rock through to arches, stacks, and stumps that can occur as erosion progresses.

Frost shattering (or freeze-thaw), caused by the expansion of frozen sea water in cavities, and biological weathering, caused by the burrowing of rock-boring molluscs, also result in the breakdown of the coastal rock.

Where resistant rocks form headlands, the sea erodes the coast in successive stages. First it exploits weaknesses, such as faults and cracks, in cave openings and then gradually wears away the interior of the caves until their roofs are pierced through to form blowholes. In time, caves at either side of a headland may unite to form a natural arch. When the roof of the arch collapses, a stack is formed. This may be worn down further to produce a stump and a wave-cut platform.

Beach erosion occurs when more sand is eroded and carried away from the beach than is deposited by longshore drift. Beach erosion can occur due to the construction of artificial barriers, such as groynes, or due to the natural periodicity of the **beach cycle**, whereby high tides and the high waves of winter storms tend to carry sand away from the beach and deposit it offshore in the form of bars. During the calmer summer season some of this sand is redeposited on the beach.

In Britain, the southern half of the coastline is slowly sinking (on the east coast, at the rate of half a centimetre a year) whilst the northern half is rising, as a result of rebounding of the land mass (responding to the removal of ice from the last Ice Age). Some areas may be eroding at a rate of 6 m/20 ft per year. Current opinion is to surrender the land to the sea, rather than build costly sea defences in rural areas. In 1996, it was reported that 29 villages had disappeared from the Yorkshire coast since 1926 as a result of tidal battering.

Glossary

coastal protection

measures taken to prevent coastal erosion. Many stretches of coastline are so severely affected by erosion that beaches are swept away, threatening the livelihood of seaside resorts, and buildings become unsafe.

To reduce erosion, several different forms of coastal protection may be employed. Structures such as sea walls attempt to prevent waves reaching the cliffs by deflecting them back to sea. Such structures are expensive and of limited success. Adding sediment (beach nourishment) to make a beach wider causes waves to break early so that they have less power when they reach the cliffs. Wooden or concrete barriers called groynes may also be constructed at right angles to the beach in order to block the movement of sand along the beach (longshore drift). This, however, has the effect of 'starving' beaches downshore: 'protection' of one area usually means destruction of another.

Coastal protection may also refer to the process of simply leaving the coast to the elements but removing the harmful factor of human population and development.

coccolithophorid

microscopic, planktonic marine alga, which secretes a calcite shell. The shells (coccoliths) of coccolithophores are a major component of deep sea ooze. Coccolithophores were particularly abundant during the late Cretaceous period and their remains form the northern European chalk deposits, such as the white cliffs of Dover.

combe or coombe

steep-sided valley found on the scarp slope of a chalk escarpment. The inclusion of 'combe' in a placename usually indicates that the underlying rock is chalk.

composite volcano or stratovolcano

steep-sided conical volcano formed above a subduction zone. It is made up of alternate layers of ash and lava. The magma (molten rock) associated with composite volcanoes is very thick and often clogs up the vent.

This can cause a tremendous buildup of pressure, which, once released, causes a very violent eruption. Examples of composite volcanoes are Mount St Helens in the USA and Mount Mayon in the Philippines. Compare shield volcano.

condensation

conversion of a vapour to a liquid. This is frequently achieved by letting the vapour come into contact with a cold surface. It is the process by which water vapour turns into fine water droplets to form cloud.

Condensation in the atmosphere occurs when the air becomes completely saturated and is unable to hold any more water vapour. As air rises it cools and contracts – the cooler it becomes the less water it can hold. Rain is frequently associated with warm weather fronts because the air rises and cools, allowing the water vapour to condense as rain. The temperature at which the air becomes saturated is known as the dew point. Water vapour will not condense in air if there are not enough condensation nuclei (particles of dust, smoke or salt) for the droplets to form on. It is then said to be supersaturated. Condensation is an important part of the water cycle.

constructive margin or divergent margin
in plate tectonics, a region in which two plates are moving away from each other. Magma, or molten rock, escapes to the surface along this margin to form new crust, usually in the form of a ridge. Over time, as more and more magma reaches the surface, the sea floor spreads – for example, the upwelling of magma at the Mid-Atlantic Ridge causes the floor of the Atlantic Ocean to grow at a rate of about 5 cm/2 in a year.

Volcanoes can form along the ridge and islands may result (for example, Iceland was formed in this way). Eruptions at constructive plate margins tend to be relatively gentle; the lava produced cools to form basalt.

continent
any one of the seven large land masses of the Earth, as distinct from the oceans. They are Asia, Africa, North America, South America, Europe, Australia, and Antarctica. Continents are constantly moving and evolving. A continent does not end at the coastline; its boundary is the edge of the shallow continental shelf, which may extend several hundred kilometres out to sea.

Continental crust, as opposed to the crust that underlies the deep oceans, is composed of a wide variety of igneous, sedimentary, and metamorphic rocks. The rocks vary in age from recent (currently forming) to almost 4000 million years old. Unlike the ocean crust, the continents are not only high standing, but extend to depths as great at 70 km/43 mi under high mountain ranges. Continents, as high, dry masses of rock, are present on Earth because of the density contrast between them and the rock that underlies the oceans. Continental crust is both thick and light, whereas ocean crust is thin and dense. If the crust were the same thickness and density everywhere, the entire Earth would be covered in water.

continental drift
theory that, about 200–250 million years ago, the Earth consisted of a single large continent (Pangaea), which subsequently broke apart to form the continents known today. The theory was proposed in 1912 by German meteorologist Alfred Wegener, but such vast continental movements could not be satisfactorily explained or even accepted by geologists until the 1960s.

The theory of continental drift gave way to the theory of of plate tectonics. Whereas Wegener proposed that continents ploughed their way through underlying mantle and ocean floor, plate tectonics states that continents are just part of larger lithospheric plates (which include ocean crust as well) that move laterally over Earth's surface.

Glossary

continental rise
portion of the ocean floor rising gently from the abyssal plain towards the steeper continental slope. The continental rise is a depositional feature formed from sediments transported down the slope mainly by turbidity currents. Much of the continental rise consists of coalescing submarine alluvial fans bordering the continental slope.

continental shelf
submerged edge of a continent, a gently sloping plain that extends into the ocean. It typically has a gradient of less than 1°. When the angle of the sea bed increases to 1–5° (usually several hundred kilometres away from land), it becomes known as the *continental slope*.

continental slope
sloping, submarine portion of a continent. It extends downward from the edge of the continental shelf. In some places, such as south of the Aleutian Islands of Alaska, continental slopes extend directly to the ocean deeps or abyssal plain. In others, such as the east coast of North America, they grade into the gentler continental rises that in turn grade into the abyssal plains.

convectional rainfall
rainfall associated with hot climates, resulting from the uprising of convection currents of warm air. Air that has been warmed by the extreme heating of the ground surface rises to great heights and is abruptly cooled. The water vapour carried by the air condenses and rain falls heavily. Convectional rainfall is often associated with a thunderstorm.

convection current
current caused by the expansion of a liquid, solid, or gas as its temperature rises. The expanded material, being less dense, rises, while colder, denser material sinks. Material of neutral buoyancy moves laterally. Convection currents arise in the atmosphere above warm land masses or seas, giving rise to sea breezes and land breezes, respectively. In some heating systems, convection currents are used to carry hot water upwards in pipes.

Convection currents in the hot, solid rock of the Earth's mantle help to drive the movement of the rigid plates making up the Earth's surface.

core
innermost part of Earth. It is divided into an outer core, which begins at a depth of 2,900 km/1,800 mi, and an inner core, which begins at a depth of 4,980 km/3,100 mi. Both parts are thought to consist of iron-nickel alloy. The outer core is liquid and the inner core is solid.

The fact that seismic shear waves disappear at the mantle–outer core boundary indicates that the outer core is molten, since shear waves cannot travel through fluid. Scientists infer the iron-nickel rich composition of the core from Earth's density and its moment of inertia, and the composition of iron meteorites, which are thought to be pieces of cores of small planets. The temperature of the core, as estimated from the melting point of iron at high pressure, is thought to be at least 4,000°C/7,200°F, but remains controversial. Earth's magnetic field is believed to be the result of the movement of liquid metal in the outer core.

Coriolis effect
effect of the Earth's rotation on the atmosphere, oceans, and theoretically all objects moving over the Earth's surface. In the northern hemisphere it causes moving objects and currents to be deflected to the right; in the southern hemisphere it causes deflection to the left. The effect is named after its discoverer, French mathematician Gaspard de Coriolis (1792–1843).

corrasion
grinding away of solid rock surfaces by particles carried by water, ice, and wind. It is generally held to be the most significant form of erosion. As the eroding particles are carried along they become eroded themselves due to the process of attrition.

corrosion
alternative name for solution, the process by which water dissolves rocks such as limestone.

corundum
native aluminium oxide, Al_2O_3, the hardest naturally occurring mineral known apart from diamond (corundum rates 9 on the Mohs scale of hardness); lack of cleavage also increases its durability. Its crystals are barrel-shaped prisms of the trigonal system. Varieties of gem-quality corundum are **ruby** (red) and **sapphire** (any colour other than red, usually blue). Poorer-quality and synthetic corundum is used in industry, for example as an abrasive.

crater
bowl-shaped depression in the ground, usually round and with steep sides. Craters are formed by explosive events such as the eruption of a volcano or the impact of a meteorite.

The Moon has more than 300,000 craters over 1 km/0.6 mi in diameter, mostly formed by meteorite bombardment; similar craters on Earth have mostly been worn away by erosion. Craters are found on all of the other rocky bodies in the Solar System.

Craters produced by impact or by volcanic activity have distinctive shapes, enabling geologists to distinguish likely methods of crater formation on planets in the Solar System. Unlike volcanic craters, impact craters have raised rims and central peaks and are circular, unless the meteorite has an extremely low angle of incidence or the crater has been affected by some later process.

Glossary

The crater of a long-dead volcano in Iceland, which has filled with water to form a crater lake. Iceland's system of volcanoes, many of them active, has been created as a result of the movement in this area of two of the Earth's tectonic plates in relation to one another.
Premaphotos Wildlife

craton or continental shield
relatively stable core of a continent that is not currently affected by tectonics along plate boundaries. Cratons generally consist of highly deformed metamorphic rock that formed during ancient orogenic explosions.

Cratons exist in the hearts of all the continents, a typical example being the Canadian Shield.

Cretaceous
period of geological time approximately 143–65 million years ago. It is the last period of the Mesozoic era, during which angiosperm (seed-bearing) plants evolved, and dinosaurs reached a peak. The end of the Cretaceous period is marked by a mass extinction of many lifeforms, most notably the dinosaurs. The north European chalk, which forms the white cliffs of Dover, was deposited during the latter half of the Cretaceous, hence the name Cretaceous, which comes from the Latin *creta*, 'chalk'.

crust
rocky outer layer of Earth, consisting of two distinct parts, the oceanic crust and the continental crust. The *oceanic* crust is on average about 10 km/6 mi thick and consists mostly of basaltic rock overlain by muddy sediments. By contrast, the *continental* crust is largely of granitic composition and is more complex in its structure. Because it is continually recycled back into the mantle

by the process of subduction, the oceanic crust is in no place older than about 200 million years. However, parts of the continental crust are over 3.5 billion years old.

Beneath a layer of surface sediment, the oceanic crust is made up of a layer of basalt, followed by a layer of gabbro. The continental crust varies in thickness from about 40 km/25 mi to 70 km/45 mi, being deepest beneath mountain ranges, and thinnest above continental rift valleys. Whereas the oceanic crust is composed almost exclusively of basaltic igneous rocks and sediments, the continental crust is made of a wide variety of sedimentary, igneous, and metamorphic rocks.

crystallography
scientific study of crystals. In 1912 it was found that the shape and size of the repeating atomic patterns (unit cells) in a crystal could be determined by passing X-rays through a sample. This method, known as X-ray diffraction, opened up an entirely new way of 'seeing' atoms. It has been found that many substances have a unit cell that exhibits all the symmetry of the whole crystal; in table salt (sodium chloride, NaCl), for instance, the unit cell is an exact cube.

Many materials were not even suspected of being crystals until they were examined by X-ray crystallography. It has been shown that purified biomolecules, such as proteins and DNA, can form crystals, and such compounds may now be studied by this method. Other applications include the study of metals and their alloys, and of rocks and soils.

current
flow of a body of water or air, or of heat, moving in a definite direction. Ocean currents are fast-flowing bodies of seawater moved by the wind or by variations in water density between two areas. They are partly responsible for transferring heat from the Equator to the poles and thereby evening out the global heat imbalance. There are three basic types of ocean current: *drift currents* are broad and slow-moving; *stream currents* are narrow and swift-moving; and *upwelling currents* bring cold, nutrient-rich water from the ocean bottom.

Stream currents include the Gulf Stream and the Japan (or Kuroshio) Current. Upwelling currents, such as the Gulf of Guinea Current and the Peru (Humboldt) Current, provide food for plankton, which in turn supports fish and sea birds.

cyclone
alternative name for a depression, an area of low atmospheric pressure with winds blowing in an anticlockwise direction in the northern hemisphere and a clockwise direction in the southern hemisphere. A severe cyclone that forms in the tropics is called a tropical cyclone or hurricane.

dating
process of determining the age of minerals, rocks, fossils, and geological formations. There are two types of dating: relative and absolute. ***Relative dating***

involves determining the relative ages of materials, that is determining the chronological order of formation of particular rocks, fossils, or formations, by means of careful field work.

Absolute dating is the process of determining the absolute age (that is, the age in years) of a mineral, rock, or fossil. Absolute dating is accomplished using methods such as radiometric dating (measuring the abundances of particular isotopes in a mineral), fission track dating, and even counting annual layers of sediment.

Deep-Sea Drilling Project
research project initiated by the USA in 1968 to sample the rocks of the ocean crust. In 1985 it became known as the Ocean Drilling Program (ODP).

deep-sea trench
another term for ocean trench.

delta
tract of land at a river's mouth, composed of silt deposited as the water slows on entering the sea. Familiar examples of large deltas are those of the Mississippi, Ganges and Brahmaputra, Rhône, Po, Danube, and Nile; the shape of the Nile delta is like the Greek capital letter *delta* Δ, and thus gave rise to the name.

The *arcuate delta* of the Nile is one form of delta. Others are *birdfoot deltas*, like that of the Mississippi which is a seaward extension of the river's levee system; and *tidal deltas*, like that of the Mekong, in which most of the material is swept to one side by sea currents.

depression or cyclone or low
in meteorology, a region of relatively low atmospheric pressure. In mid latitudes a depression forms as warm, moist air from the tropics mixes with cold, dry polar air, producing warm and cold boundaries (fronts) and unstable weather – low cloud and drizzle, showers, or fierce storms. The warm air, being less dense, rises above the cold air to produce the area of low pressure on the ground. Air spirals in towards the centre of the depression in an anticlockwise direction in the northern hemisphere, clockwise in the southern hemisphere, generating winds up to gale force. Depressions tend to travel eastwards and can remain active for several days.

A deep depression is one in which the pressure at the centre is very much lower than that round about; it produces very strong winds, as opposed to a shallow depression, in which the winds are comparatively light. A severe depression in the tropics is called a hurricane, tropical *cyclone*, or typhoon, and is a great danger to shipping; a tornado is a very intense, rapidly swirling depression, with a diameter of only a few hundred metres or so.

destructive margin
in plate tectonics, a region on the Earth's crust in which two plates are moving towards one another. Usually one plate (the denser of the two) is forced to dive

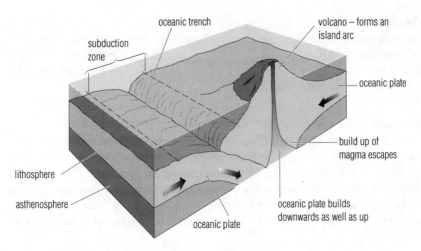

When crustal plates meet, and one plate is denser than the other, the denser plate is forced under the other plate (at the subduction zone) and melts to form magma. If both plates are of equal density they collide and crumple up against each other forming mountains.

below the other into what is called the *subduction zone*. The descending plate melts to form a body of magma, which may then rise to the surface through cracks and faults to form volcanoes. If the two plates consist of more buoyant continental crust, subduction does not occur. Instead, the crust crumples gradually to form fold mountains, such as the Himalayas.

Devonian period
period of geological time 408–360 million years ago, the fourth period of the Palaeozoic era. Many desert sandstones from North America and Europe date from this time. The first land plants flourished in the Devonian period, corals were abundant in the seas, amphibians evolved from air-breathing fish, and insects developed on land.

The name comes from the county of Devon in southwest England, where Devonian rocks were first studied.

dew point
temperature at which the air becomes saturated with water vapour. At temperatures below the dew point, the water vapour condenses out of the air as droplets. If the droplets are large they become deposited on the ground as dew; if small they remain in suspension in the air and form mist or fog.

diagenesis
physical, chemical, and biological processes by which a sediment becomes a sedimentary rock. The main processes involved include compaction of the grains, and the cementing of the grains together by the growth of new minerals deposited by percolating groundwater. As a whole, diagenesis is actually a poorly understood process.

diamond
generally colourless, transparent mineral, an allotrope of carbon. It is regarded as a precious gemstone, and is the hardest substance known (10 on the Mohs scale). Industrial diamonds, which may be natural or synthetic, are used for cutting, grinding, and polishing.

diamond-anvil cell
device composed of two opposing cone-shaped diamonds that when squeezed together by a lever-arm exert extreme pressures. The high pressures result from applying force to the small areas of the opposing diamond faces. The device is used to determine the properties of materials at pressures corresponding to those of planetary interiors. One discovery made with the diamond-anvil cell is $MgSiO_3$-perovskite, thought to be the predominant mineral of Earth's lower mantle.

diapirism
geological process in which a particularly light rock, such as rock salt, punches upwards through the heavier layers above. The resulting structure is called a salt dome, and oil is often trapped in the curled-up rocks at each side.

dip
angle at which a structural surface, such as a fault or a bedding plane, is inclined from horizontal. Measured at right angles to the strike of that surface, it is used together with strike to describe the orientation of geological features in the field. Rocks that are dipping have usually been affected by folding.

divergent margin
in plate tectonics, the boundary or active zone between two lithospheric plates that are moving away from each other. Divergent margins between plates of oceanic lithosphere are characterized by volcanic ocean ridges. Divergent margins within a continent are usually less well-defined, and are characterized by volcanically active continental rift valleys (for example the East African rift valley).

Volcanoes can form along the ridge and islands may result (for example, Iceland was formed in this way). Eruptions at divergent plate margins tend to be relatively gentle; the lava produced cools to form basalt.

doldrums
area of low atmospheric pressure along the Equator, in the intertropical convergence zone where the northeast and southeast trade winds converge. The doldrums are characterized by calm or very light winds, during which there may be sudden squalls and stormy weather. For this reason the areas are avoided as far as possible by sailing ships.

dolomite

white mineral with a rhombohedral structure, calcium magnesium carbonate $(CaMg(CO_3)_2)$. Dolomites are common in geological successions of all ages and are often formed when limestone is changed by the replacement of the mineral calcite with the mineral dolomite.

drumlin

long streamlined hill created in formerly glaciated areas. Rocky debris (till) is gathered up by the glacial icesheet and moulded to form an egg-shaped mound, 8–60 m/25–200 ft in height and 0.5–1 km/0.3–0.6 mi in length. Drumlins commonly occur in groups on the floors of glacial troughs, producing a 'basket-of-eggs' landscape.

They are important indicators of the direction of ice flow, as their blunt ends point upstream, and their gentler slopes trail off downstream.

dune

mound or ridge of wind-drifted sand common on coasts and in deserts. Loose sand is blown and bounced along by the wind, up the windward side of a dune. The sand particles then fall to rest on the lee side, while more are blown up from the windward side. In this way a dune moves gradually downwind.

In sandy deserts, the typical crescent-shaped dune is called a *barchan*. *Seif dunes* are longitudinal and lie parallel to the wind direction, and *star-shaped dunes* are formed by irregular winds.

dyke

sheet of igneous rock created by the intrusion of magma (molten rock) across layers of pre-existing rock. (By contrast, a sill is intruded *between* layers of rock.) It may form a ridge when exposed on the surface if it is more resistant than the rock into which it intruded. A dyke is also a human-made embankment built along a coastline (for example, in the Netherlands) to prevent the flooding of lowland coastal regions.

Earth

third planet from the Sun. It is almost spherical, flattened slightly at the poles, and is composed of five concentric layers: inner core, outer core, mantle, crust, and atmosphere. About 70% of the surface (including the north and south polar icecaps) is covered with water. The Earth is surrounded by a life-supporting atmosphere and is the only planet on which life is known to exist.

mean distance from the Sun
149,500,000 km/92,860,000 mi

equatorial diameter
12,755 km/7,920 mi

circumference
40,070 km/24,900 mi

Glossary

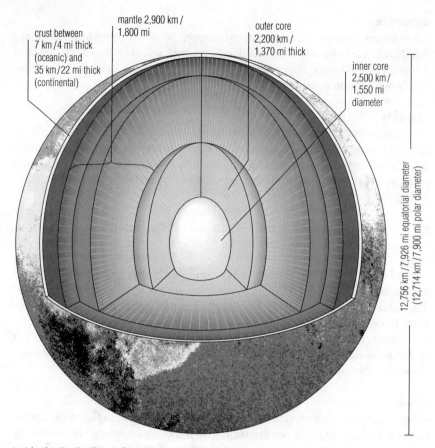

Inside the Earth. The surface of the Earth is a thin crust about 6 km/4 mi thick under the sea and 40 km/25 mi thick under the continents. Under the crust lies the mantle about 2,900 km/1,800 mi thick and with a temperature of 1,500–3,000°C/2,700–5,400°F. The outer core is about 2,250 km/1,400 mi thick, of molten iron and nickel. The inner core is probably solid iron and nickel at about 5,000°C/9,000°F.

rotation period
23 hr 56 min 4.1 sec

year
(complete orbit, or sidereal period) 365 days 5 hr 48 min 46 sec. Earth's average speed around the Sun is 30 kps/18.5 mps; the plane of its orbit is inclined to its equatorial plane at an angle of 23.5°, the reason for the changing seasons

atmosphere
nitrogen 78.09%; oxygen 20.95%; argon 0.93%; carbon dioxide 0.03%; and less than 0.0001% neon, helium, krypton, hydrogen, xenon, ozone, radon

surface

land surface 150,000,000 sq km/57,500,000 sq mi (greatest height above sea level 8,872 m/29,118 ft Mount Everest); water surface 361,000,000 sq km/139,400,000 sq mi (greatest depth 11,034 m/36,201 ft Mariana Trench in the Pacific). The interior is thought to be an inner core about 2,600 km/1,600 mi in diameter, of solid iron and nickel; an outer core about 2,250 km/1,400 mi thick, of molten iron and nickel; and a mantle of mostly solid rock about 2,900 km/1,800 mi thick. The crust and the uppermost layer of the mantle form about 12 major moving plates, some of which carry the continents. The plates are in constant, slow motion, called tectonic drift

satellite
the Moon

age
4.6 billion years. The Earth was formed with the rest of the Solar System by consolidation of interstellar dust. Life began 3.5–4 billion years ago.

earthquake
abrupt motion that propagates through the Earth and along its surfaces. Earthquakes are caused by the sudden release in rocks of strain accumulated over time as a result of tectonics. The study of earthquakes is called seismology. Most earthquakes occur along faults (fractures or breaks) and Benioff zones. Plate tectonic movements generate the major proportion: as two plates move passed each other they can become jammed. When sufficient strain has accumulated, the rock breaks, releasing a series of elastic waves (seismic waves) as the plates spring free. The force of earthquakes (magnitude) is measured on the Richter scale, and their effect (intensity) on the Mercalli scale. The point at which an earthquake originates is the *seismic focus* or *hypocentre*; the point on the Earth's surface directly above this is the *epicentre*.

earth science
scientific study of the planet Earth as a whole. The mining and extraction of minerals and gems, the prediction of weather and earthquakes, the pollution of the atmosphere, and the forces that shape the physical world all fall within its scope of study. The emergence of the discipline reflects scientists' concern that an understanding of the global aspects of the Earth's structure and its past will hold the key to how humans affect its future, ensuring that its resources are used in a sustainable way. It is a synthesis of several traditional subjects such as geology, meteorology, oceanography, geophysics, geochemistry, and palaeontology.

Ekman spiral effect
in oceanography, theoretical description of a consequence of the Coriolis effect on ocean currents, whereby currents flow at an angle to the winds that drive them. It derives its name from the Swedish oceanographer Vagn Ekman (1874–1954).

In the northern hemisphere, surface currents are deflected to the right of the wind direction. The surface current then drives the subsurface layer at an angle to its original deflection. Consequent subsurface layers are similarly affected, so that the effect decreases with increasing depth. The result is that most water is transported at about right-angles to the wind direction. Directions are reversed in the southern hemisphere.

electromagnetic radiation
transfer of energy in the form of electromagnetic waves.

electromagnetic spectrum
complete range, over all wavelengths and frequencies, of electromagnetic waves. These include radio and television waves, infrared radiation, visible light, ultraviolet light, X-rays, and gamma radiation.

electromagnetic waves
oscillating electric and magnetic fields travelling together through space at a speed of nearly 300,000 km/186,000 mi per second. The (limitless) range of possible wavelengths and frequencies of electromagnetic waves, which can be thought of as making up the *electromagnetic spectrum*, includes radio waves, infrared radiation, visible light, ultraviolet radiation, X-rays, and gamma rays.

Radio and television waves lie at the *long wavelength–low frequency* end of the spectrum, with wavelengths longer than 10^{-4} m. Infrared radiation has wavelengths between 10^{-4} m and 7×10^{-7} m. Visible light has yet shorter wavelengths from 7×10^{-7} m to 4×10^{-7} m. Ultraviolet radiation is near the *short wavelength–high frequency* end of the spectrum, with wavelengths between 4×10^{-7} m and 10^{-8} m. X-rays have wavelengths from 10^{-8} to 10^{-12}. Gamma radiation has the shortest wavelengths of less than 10^{-10}. The different wavelengths and frequencies lend specific properties to electromagnetic waves. While visible light is diffracted by a diffraction grating, X-rays can only be diffracted by crystals. Radio waves are refracted by the atmosphere; visible light is refracted by glass or water.

El Niño
oceanographic and meteorological condition that occurs when a warm, nutrient-poor current of water moves southward to replace the cold, nutrient-rich surface water of the Peru Current along the west coast of South America. The phenomenon occurs every year around Christmas and lasts for several weeks, but in some years can lasts for several months. The result is poor fish harvests and global changes in weather patterns.

emerald
clear, green gemstone variety of the mineral beryl. It occurs naturally in Colombia, the Ural Mountains in Russia, Zimbabwe, and Australia. The green colour is caused by the presence of the element chromium in the beryl.

emery
black to greyish form of impure corundum that also contains the minerals magnetite and haematite. It is used as an abrasive.

environmental impact assessment (EIA)
in the UK, a process by which the potential environmental impacts of human activities, such as the construction of a power station, dam, or major housing development, are evaluated. The results of an EIA are published and discussed by different levels of government, non-governmental organizations, and the general public before a decision is made on whether or not the project can proceed.

Some developments, notably those relating to national defence, are exempt from EIA. Increasingly studies include the impact not only on the physical environment, but also the socio-economic environment, such as the labour market and housing supply.

environmental issues
matters relating to the detrimental effects of human activity on the biosphere, their causes, and the search for possible solutions. Since the Industrial Revolution, the demands made by both the industrialized and developing nations on the Earth's natural resources are increasingly affecting the balance of the Earth's resources. Over a period of time, some of these resources are renewable – trees can be replanted, soil nutrients can be replenished – but many resources, such as fossil fuels and minerals, are non-renewable and in danger of eventual exhaustion. In addition, humans are creating many other problems that may endanger not only their own survival, but also that of other species. For instance, deforestation and air pollution are not only damaging and radically altering many natural environments, they are also affecting the Earth's climate by adding to the greenhouse effect and global warming, while water pollution is seriously affecting aquatic life, including fish populations, as well as human health.

Eocene epoch
second epoch of the Tertiary period of geological time, roughly 56.5–35.5 million years ago. Originally considered the earliest division of the Tertiary, the name means 'early recent', referring to the early forms of mammals evolving at the time, following the extinction of the dinosaurs.

epicentre
point on the Earth's surface immediately above the seismic focus of an earthquake. Most building damage usually takes place at an earthquake's epicentre. The term sometimes refers to a point directly above or below a nuclear explosion ('at ground zero').

epoch
subdivision of a geological period in the geological time scale. Epochs are sometimes given their own names (such as the Palaeocene, Eocene, Oligocene,

Miocene, and Pliocene epochs comprising the Tertiary period), or they are referred to as the late, early, or middle portions of a given period (as the Late Cretaceous or the Middle Triassic epoch).

Geological time is broken up into **geochronological units** of which epoch is just one level of division. The hierarchy of geochronological divisions is eon, era, period, epoch, age, and chron. Epochs are subdivisions of periods and ages are subdivisions of epochs. Rocks representing an epoch of geological time comprise a *series*.

era
any of the major divisions of geological time that includes several periods but is part of an eon. The eras of the current Phanerozoic in chronological order are the Palaeozoic, Mesozoic, and Cenozoic. We are living in the recent epoch of the Quaternary period of the Cenozoic era.

Geological time is broken up into **geochronological units** of which era is just one level of division. The hierarchy of geochronological divisions is eon, era, period, epoch, age, and chron. Eras are subdivisions of eons and periods are subdivisions of eras. Rocks representing an era of geological time comprise an *erathem*.

erosion
wearing away of the Earth's surface, caused by the breakdown and transportation of particles of rock or soil (by contrast, weathering does not involve transportation). Agents of erosion include the sea, rivers, glaciers, and wind.

Water, consisting of sea waves and currents, rivers, and rain; ice, in the form of glaciers; and wind, hurling sand fragments against exposed rocks and moving dunes along, are the most potent forces of erosion.

People also contribute to erosion by bad farming practices and the cutting down of forests, which can lead to the formation of dust bowls.

There are several processes of erosion including hydraulic action, corrasion, attrition, and solution.

erratic
displaced rock that has been transported by a glacier or some other natural force to a site of different geological composition.

For example, in East Anglia, England, erratics have been found that have been transported from as far away as Scotland and Scandinavia.

escarpment or cuesta
large ridge created by the erosion of dipping sedimentary rocks. It has one steep side (scarp) and one gently sloping side (dip). Escarpments are common features of chalk landscapes, such as the Chiltern Hills and the North Downs in England. Certain features are associated with chalk escarpments, including dry valleys (formed on the dip slope), combes (steep-sided valleys on the scarp slope), and springs.

esker
narrow, steep-walled ridge, often sinuous and sometimes branching, formed beneath a glacier. It is made of sands and gravels, and represents the course of a subglacial river channel. Eskers vary in height from 3–30 m/10–100 ft and can be up to 160 km/100 mi or so in length.

Etesian wind
north-northwesterly wind that blows June–September in the eastern Mediterranean and Aegean seas.

eustatic change
global rise or fall in sea level caused by a change in the amount of water in the oceans (by contrast, isostatic adjustment involves a rising or sinking of the land, which causes a local change in sea level). During the last ice age, global sea level was lower than today because water became 'locked-up' in the form of ice and snow, and less water reached the oceans.

exosphere
uppermost layer of the atmosphere. It is an ill-defined zone above the thermosphere, beginning at about 700 km/435 mi and fading off into the vacuum of space. The gases are extremely thin, with hydrogen as the main constituent.

extrusive rock or volcanic rock
igneous rock that solidifies on the surface of the Earth (as opposed to intrusive, or plutonic, rocks that solidify below Earth's surface). Most extrusive rocks are finely grained because they crystallize so quickly that the crystals do not have time to grow very large. Extrusive rocks include those that formed from lava flowing out of a volcano (for example basalt), and those made of welded fragments of ash and glass that fell from the sky after being ejected into the air during a volcanic eruption.

facies
body of rock strata possessing unifying characteristics usually indicative of the environment in which the rocks were formed. The term is also used to describe the environment of formation itself or unifying features of the rocks that comprise the facies.

Features that define a facies can include collections of fossils, sequences of rock layers, or the occurrence of specific minerals. Sedimentary rocks deposited at the same time, but representing different facies belong to a single *chronostratigraphic unit*. But these same rocks may belong to different *lithostratigraphic units*. For example, beach sand is deposited at the same time that mud is deposited further offshore. The beach sand eventually turns to sandstone while the mud turns to shale. The resulting sandstone and shale *strata* comprise two different facies, one representing the beach environment and the other the offshore environment, formed at the same time; the sandstone and shale belong to the same chronostratigraphic unit but distinct lithostratigraphic units.

fault

planar break in rocks, along which the rock formations on either side have moved relative to one another. Faults involve displacements, or offsets, ranging from the microscopic scale to hundreds of kilometres. Large offsets along a fault are the result of the accumulation of smaller movements (metres or less) over long periods of time. Large motions cause detectable earthquakes.

Faults are planar features. Fault orientation is described by the inclination of the fault plane with respect to horizontal (see dip) and its direction in the horizontal plane. Faults at high angle with respect to horizontal (in which the fault plane is steep) are classified as either ***normal faults***, where the hanging wall (the body of rock above the fault) has moved down relative to the footwall (the body of rock below the fault), or ***reverse faults***, where the hanging wall has moved up relative to the footwall. Normal faults occur where rocks on either side have moved apart. Reverse faults occur where rocks on either side have been forced together. A reverse fault that forms a low angle with the horizontal plane is called a ***thrust fault***.

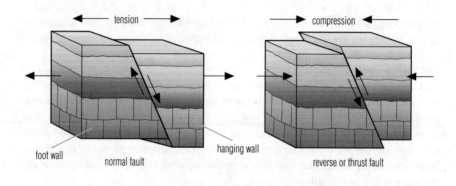

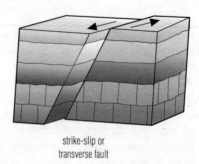

Faults are caused by the movement of rock layers, producing such features as block mountains and rift valleys. A normal fault is caused by a tension or stretching force acting in the rock layers; a reverse fault is caused by compression forces; and a strike-slip fault is caused by a sideways force. Faults can continue to move for thousands or millions of years.

A *lateral fault*, or *strike–slip fault*, occurs where the relative movement along the fault plane is sideways. A *transform fault* is a major strike–slip fault along a plate boundary, that joins two other plate boundaries – two spreading centres, two subduction zones, or one spreading centre and one subduction zone. The San Adreas fault is a transform fault.

Faults produce lines of weakness on the Earth's surface (along their strike) that are often exploited by processes of weathering and erosion. Coastal caves and geos (narrow inlets) often form along faults and, on a larger scale, rivers may follow the line of a fault.

feldspar

group of silicate minerals. Feldspars are the most abundant mineral type in the Earth's crust. They are the chief constituents of igneous rock and are present in most metamorphic and sedimentary rocks. All feldspars contain silicon, aluminium, and oxygen, linked together to form a framework. Spaces within this framework structure are occupied by sodium, potassium, calcium, or occasionally barium, in various proportions. Feldspars form white, grey, or pink crystals and rank 6 on the Mohs scale of hardness.

felsic rock

plutonic rock composed chiefly of light-coloured minerals, such as quartz, feldspar and mica. It is derived from *feldspar*, *lenad* (meaning feldspathoid), and *silica*. The term *felsic* also applies to light-coloured minerals as a group, especially quartz, feldspar, and feldspathoids.

fjord or fiord

narrow sea inlet enclosed by high cliffs. Fjords are found in Norway, New Zealand, and western parts of Scotland. They are formed when an overdeepened U-shaped glacial valley is drowned by a rise in sea-level. At the mouth of the fjord there is a characteristic lip causing a shallowing of the water. This is due to reduced glacial erosion and the deposition of moraine at this point.

Fiordland is the deeply indented southwest coast of South Island, New Zealand; one of the most beautiful inlets is Milford Sound.

flint

compact, hard, brittle mineral (a variety of chert), brown, black, or grey in colour, found as nodules in limestone or shale deposits. It consists of cryptocrystalline (grains too small to be visible even under a light microscope) silica, SiO_2, principally in the crystalline form of quartz. Implements fashioned from flint were widely used in prehistory.

flood plain

area of periodic flooding along the course of river valleys. When river discharge exceeds the capacity of the channel, water rises over the channel banks and floods the adjacent low-lying lands. As water spills out of the channel some alluvium (silty material) will be deposited on the banks to form levees (raised river banks). This water will slowly seep into the flood plain, depositing a new layer of rich fertile alluvium as it does so.

Many important floodplains, such as the inner Niger delta in Mali, occur in arid areas where their exceptional productivity has great importance for the local economy.

A flood plain (sometimes called inner delta) can be regarded as part of a river's natural domain, statistically certain to be claimed by the river at repeated intervals. By plotting floods that have occurred and extrapolating from these data we can speak of 10-year floods, 100-year floods, 500-year floods, and so forth, based on the statistical probability of flooding across certain parts of the flood plain.

Even the most energetic flood-control plans (such as dams, dredging, and channel modification) will sometimes fail, and using flood plains as the site of towns and villages is always laden with risk. It is more judicious to use flood plains in ways compatible with flooding, such as for agriculture or parks.

Flood plain features include meanders and oxbow lakes.

fluvial

of or pertaining to streams or rivers. A *fluvial deposit* is sedimentary material laid down by a stream or river, such as a sandstone or conglomerate (coarse-grained clastic sedimentary rock composed of rounded pebbles of pre-existing rock cemented in a fine-grained sand or clay matrix).

fluvioglacial

of a process or landform, associated with glacial meltwater. Meltwater, flowing beneath or ahead of a glacier, is capable of transporting rocky material and creating a variety of landscape features, including eskers, kames, and outwash plains.

focus

point within the Earth's crust at which an earthquake originates. The point on the surface that is immediately above the focus is called the epicentre.

fold

bend in beds or layers of rock. If the bend is arched up in the middle it is called an *anticline*; if it sags downwards in the middle it is called a *syncline*. The line along which a bed of rock folds is called its axis. The axial plane is the plane joining the axes of successive beds.

fold mountain

obsolete term used to refer to mountains formed at a convergent margin. See mountain.

foredeep

elongated structural basin lying inland from an active mountain system and receiving sediment from the rising mountains. According to plate tectonic

Glossary

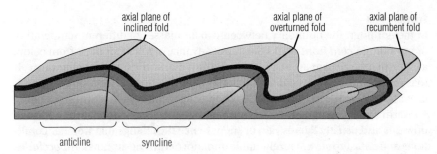

The folding of rock strata occurs where compression causes them to buckle. Over time, folding can assume highly complicated forms, as can sometimes be seen in the rock layers of cliff faces or deep cuttings in the rock. Folding contributed to the formation of great mountain chains such as the Himalayas.

theory, a mountain chain forming behind a subduction zone along a continental margin develops a foredeep or gently sloping trough parallel to it on the landward side. Foredeeps form rapidly and are usually so deep initially that the sea floods them through gaps in the mountain range. As the mountain system evolves, sediments choke the foredeep, pushing out marine water. As marine sedimentation stops, only nonmarine deposits from the rapidly eroding mountains are formed. These consist of alluvial fans and also rivers, flood plains, and related environments inland.

Before the advent of plate tectonic theory, such foredeep deposits and changes in sediments had been interpreted as sedimentary troughs, called geosynclines, that were supposed ultimately to build upward into mountains.

fossil

cast, impression, or the actual remains of an animal or plant preserved in rock. Fossils were created during periods of rock formation, caused by the gradual accumulation of sediment over millions of years at the bottom of the sea bed or an inland lake. Fossils may include footprints, an internal cast, or external impression. A few fossils are preserved intact, as with mammoths fossilized in Siberian ice, or insects trapped in tree resin that is today amber. The study of fossils is called palaeontology. Palaeontologists are able to deduce much of the geological history of a region from fossil remains.

freeze–thaw

form of physical weathering, common in mountains and glacial environments, caused by the expansion of water as it freezes. Water in a crack freezes and expands in volume by 9% as it turns to ice. This expansion exerts great pressure on the rock causing the crack to enlarge. After many cycles of freeze–thaw, rock fragments may break off to form scree slopes.

For freeze–thaw to operate effectively the temperature must fluctuate regularly above and below 0°C/32°F. It is therefore uncommon in areas of extreme and perpetual cold, such as the polar regions.

front

in meteorology, the boundary between two air masses of different temperature or humidity. A *cold front* marks the line of advance of a cold air mass from below, as it displaces a warm air mass; a *warm front* marks the advance of a warm air mass as it rises up over a cold one. Frontal systems define the weather of the mid-latitudes, where warm tropical air is continually meeting cold air from the poles.

Warm air, being lighter, tends to rise above the cold; its moisture is carried upwards and usually falls as rain or snow, hence the changeable weather conditions at fronts. Fronts are rarely stable and move with the air mass. An *occluded front* is a composite form, where a cold front catches up with a warm front and merges with it.

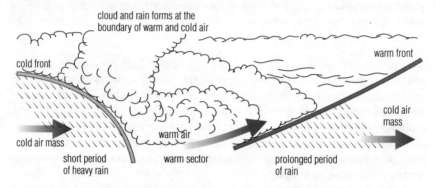

The boundaries between two air masses of different temperature and humidity. A warm front is when warm air displaces cold air; if cold air replaces warm air, it is a cold front.

gabbro

mafic (consisting primarily of dark-coloured crystals) igneous rock formed deep in the Earth's crust. It contains pyroxene and calcium-rich feldspar, and may contain small amounts of olivine and amphibole. Its coarse crystals of dull minerals give it a speckled appearance.

garnet

group of silicate minerals with the formula $X_3Y_3(SiO_4)_3$, where X is calcium, magnesium, iron, or manganese, and Y is usually aluminium or sometimes iron or chromium. Garnets are used as semiprecious gems (usually pink to deep red) and as abrasives. They occur in metamorphic rocks such as gneiss and schist.

gelifluction

type of solifluction (downhill movement of water-saturated topsoil) associated with frozen ground.

gem

mineral valuable by virtue of its durability (hardness), rarity, and beauty, cut and polished for ornamental use, or engraved. Of 120 minerals known to have been used as gemstones, only about 25 are in common use in jewellery today; of these, the diamond, emerald, ruby, and sapphire are classified as precious, and all the others semiprecious; for example, the topaz, amethyst, opal, and aquamarine.

geochemistry
science of chemistry as it applies to geology. It deals with the relative and absolute abundances of the chemical elements and their isotopes in the Earth, and also with the chemical changes that accompany geologic processes.

geochronology
branch of earth science that deals with the dating of rocks, minerals, and fossils in order to create an accurate and precise geological history of the Earth. The geological time scale is a result of these studies. It puts stratigraphic units in chronological order and assigns actual dates, in millions of years, to those units.

geodesy
science of measuring and mapping Earth's surface for making maps and correlating geological, gravitational, and magnetic measurements. Geodetic surveys, formerly carried out by means of various measuring techniques on the surface, are now commonly made by using radio signals and laser beams from orbiting satellites.

geography
study of the Earth's surface; its topography, climate, and physical conditions, and how these factors affect people and society. It is usually divided into *physical geography*, dealing with landforms and climates, and *human geography*, dealing with the distribution and activities of peoples on Earth.

geological time
time scale embracing the history of the Earth from its physical origin to the present day. Geological time is traditionally divided into eons (Archaean or Archaeozoic, Proterozoic, and Phanerozoic in ascending chronological order), which in turn are subdivided into eras, periods, epochs, ages, and finally chrons.

The terms eon, era, period, epoch, age and chron are *geochronological units* representing intervals of geological time. Rocks representing an interval of geological time comprise a *chronostratigraphic* (or *time-stratigraphic*) *unit*. Each of the hierarchical geochronological terms has a chronostratigraphic equivalent. Thus, rocks formed during an eon (a geochronological unit) are members of an eonothem (the chronostratigraphic unit equivalent of eon). Rocks of an era belong to an erathem. The chronostratigraphic equivalents of period, epoch, age, and chron are system, series, stage, and chronozone, repectively.

geology
science of the Earth, its origin, composition, structure, and history. It is divided into several branches, inlcuding *mineralogy* (the minerals of Earth), *petrology* (rocks), *stratigraphy* (the deposition of successive beds of sedimentary rocks), *palaeontology* (fossils), and *tectonics* (the deformation and movement of the Earth's crust), *geophysics* (using physics to study the Earth's surface, interior, and atmosphere), and *geochemistry* (the science of chemistry as it applies to geology).

geomorphology
branch of earth science, developed in the late 19th century, dealing with the morphology, or form, of the Earth's surface; nowadays it is also considered to be an integral part of physical geography.

Geomorphological studies involve investigating the nature and origin of surface landforms, such as mountains, valleys, plains, and plateaux, and the processes that influence them. These processes include the effects of tectonic forces, weathering, running water, waves, glacial ice, and wind, which result in the erosion, transportation, and deposition of rocks and soils. The underlying dynamics of these forces are the energy derived from the Earth's gravitational field, the flow of solar energy through the hydrological cycle, and the flow of heat from the Earth's molten interior. The mechanisms of these processes are both destructive and constructive; out of the destruction or modification of one landform another will be created.

In addition to the natural processes that mould landforms, human activity can produce changes, either directly or indirectly, and cause the erosion, transportation, and deposition of rocks and soils – for example, by poor land management practices and techniques in farming and forestry, and in the mining and construction industries. Geomorphology deals with changes in landforms over seconds or eons, and in spatial scales ranging from undulations to mountains.

geophysics
branch of earth science using physics (for instance gravity, seismicity, and magnetism) to study the Earth's surface, interior, and atmosphere. Geophysics includes several sub-fields such as seismology, palaeomagnetism, and remote sensing.

geyser
natural spring that intermittently discharges an explosive column of steam and hot water into the air due to the build-up of steam in underground chambers. One of the most remarkable geysers is Old Faithful, in Yellowstone National Park, Wyoming, USA. Geysers also occur in New Zealand and Iceland.

glacial erosion
wearing-down and removal of rocks and soil by a glacier. Glacial erosion forms impressive landscape features, including glacial troughs (U-shaped valleys),

arêtes (steep ridges), corries (enlarged hollows), and pyramidal peaks (high mountain peaks with concave faces).

Erosional landforms result from abrasion and plucking of the underlying bedrock. Abrasion is caused by the lodging of rock debris in the sole of the glacier, followed by friction and wearing away of the bedrock as the ice moves. The action is similar to that of sandpaper attached to a block of wood. The results include the polishing and scratching of rock surfaces to form powdered rock flour, and sub-parallel scratches or striations which indicate the direction of ice movement. Plucking is a form of glacial erosion restricted to the lifting and removal of blocks of bedrock already loosened by freeze-thaw activity in joint fracture.

The most extensive period of recent glacial erosion was the Pleistocene epoch in the Quaternary period when, over 2 to 3 million years, the polar ice caps repeatedly advanced and retreated. More ancient glacial episodes are also preserved in the geological record, the earliest being in the middle Precambrian and the most extensive in Permo-Carboniferous times.

glacial trough or U-shaped valley

steep-sided, flat-bottomed valley formed by a glacier. The erosive action of the glacier and of the debris carried by it results in the formation not only of the trough itself but also of a number of associated features, such as truncated spurs (projections of rock that have been sheared off by the ice) and hanging valleys (smaller glacial valleys that enter the trough at a higher level than the trough floor). Features characteristic of glacial deposition, such as drumlins and eskers, are commonly found on the floor of the trough, together with linear lakes called ribbon lakes.

gneiss

coarse-grained metamorphic rock, formed under conditions of high temperature and pressure, and often occurring in association with schists and granites. It has a foliated, or layered, structure consisting of thin bands of micas and/or amphiboles dark in colour alternating with bands of granular quartz and feldspar that are light in colour. Gneisses are formed during regional metamorphism; **paragneisses** are derived from metamorphism of sedimentary rocks and **orthogneisses** from metamorphism of granite or similar igneous rocks.

Gondwanaland or Gondwana

southern landmass formed 200 million years ago by the splitting of the single world continent Pangaea. (The northern landmass was Laurasia.) It later fragmented into the continents of South America, Africa, Australia, and Antarctica, which then drifted slowly to their present positions. The baobab tree found in both Africa and Australia is a relic of this ancient landmass.

A database of the entire geology of Gondwanaland has been constructed by geologists in South Africa. The database, known as Gondwana Geoscientific Indexing Database (GO-GEOID), displays information as a map of Gondwana 155 million years ago, before the continents drifted apart.

Glossary

granite
coarse-grained intrusive igneous rock, typically consisting of the minerals quartz, feldspar, and biotite mica. It may be pink or grey, depending on the composition of the feldspar. Granites are chiefly used as building materials.

graphite
blackish-grey, laminar, crystalline form of carbon. It is used as a lubricant and as the active component of pencil lead.

gravel
coarse sediment consisting of pebbles or small fragments of rock, originating in the beds of lakes and streams or on beaches. Gravel is quarried for use in road building, railway ballast, and for an aggregate in concrete. It is obtained from quarries known as gravel pits, where it is often found mixed with sand or clay.

gravimetry
measurement of the Earth's gravitational field. Small variations in the gravitational field (gravimetric anomalies) can be caused by varying densities of rocks and structure beneath the surface. Such variations are measured by a device called a gravimeter (or gravity-meter), which consists of a weighted spring that is pulled further downwards where the gravity is stronger. Gravimetry is used by geologists to map the subsurface features of the Earth's crust, such as underground masses of dense rock such as iron ore, or light rock such as salt.

gravity
force of attraction that arises between objects by virtue of their masses. On Earth, gravity is the force of attraction between any object in the Earth's gravitational field and the Earth itself. It is regarded as one of the four fundamental forces of nature, the other three being the electromagnetic force, the strong nuclear force, and the weak nuclear force. The gravitational force is the weakest of the four forces, but it acts over great distances. The particle that is postulated as the carrier of the gravitational force is the graviton.

One of the earliest gravitational experiments was undertaken by Nevil Maskelyne in 1774 and involved the measurement of the attraction of Mount Schiehallion (Scotland) on a plumb bob.

great circle
circle drawn on a sphere such that the diameter of the circle is a diameter of the sphere. On the Earth, all meridians of longitude are half great circles; among the parallels of latitude, only the Equator is a great circle.

The shortest route between two points on the Earth's surface is along the arc of a great circle. These are used extensively as air routes although on maps, owing to the distortion brought about by projection, they do not appear as straight lines.

greenhouse effect

phenomenon of the Earth's atmosphere by which solar radiation, trapped by the Earth and re-emitted from the surface as infrared radiation, is prevented from escaping by various gases in the atmosphere. Greenhouse gases trap heat because they readily absorb infrared radiation. The result of the greenhouse effect is a rise in the Earth's temperature (global warming). The main greenhouse gases are carbon dioxide, methane, and chlorofluorocarbons (CFCs) as well as water vapour. Fossil-fuel consumption and forest fires are the principal causes of carbon dioxide build-up; methane is a byproduct of agriculture (rice, cattle, sheep).

groundwater

water present underground in porous rock strata and soils; it emerges at the surface as springs and streams. The groundwater's upper level is called the *water table*. Rock strata that are filled with groundwater that can be extracted are called *aquifers*. Aquifers must be both porous (filled with holes) and permeable – full of holes that are interconnected so that the water is able to flow.

Most groundwater near the surface moves slowly through the ground while the water table stays in the same place. The depth of the water table reflects the balance between the rate of infiltration, called recharge, and the rate of discharge at springs or rivers or pumped water wells. The force of gravity makes underground water run 'downhill' underground just as it does above the surface. The greater the slope and the permeability, the greater the speed. Velocities vary from 100 cm/40 in per day to 0.5 cm/0.2 in.

groyne

wooden or concrete barrier built at right angles to a beach in order to block the movement of material along the beach by longshore drift. Groynes are usually successful in protecting individual beaches, but because they prevent beach material from passing along the coast they can mean that other beaches, starved of sand and shingle, are in danger of being eroded away by the waves. This happened, for example, at Barton-on-Sea in Hampshire, England, in the 1970s, following the construction of a large groyne at Bournemouth.

gryke

enlarged joint that separates blocks of limestone (clints) in a limestone pavement.

Gulf Stream

warm ocean current that flows north from the warm waters of the Gulf of Mexico along the east coast of America, from which it is separated by a channel of cold water originating in the southerly Labrador current. Off Newfoundland, part of the current is diverted east across the Atlantic, where it is known as the *North Atlantic Drift*, dividing to flow north and south, and warming what would otherwise be a colder climate in the British Isles and northwest Europe.

At its beginning the Gulf Stream is 80–150 km/50–93 mi wide and up to 850 m/2,788 ft deep, and moves with an average velocity of 130 km/80 mi a day. Its temperature is about 26°C/78°F. As it flows northwards, the current cools and becomes broader and less rapid.

guyot
flat-topped seamount. Such undersea mountains are found throughout the abyssal plains of major ocean basins, and most of them are covered by an appreciable depth of water, sediment, and ancient coral. They are believed to have started as volcanic cones formed near mid-oceanic ridges or other hot spots, in relatively shallow water, and to have been truncated by wave action as their tops emerged above the surface. As they are transported away from the ridge or other birthplace, the ocean crust ages, cools, and sinks along with the seamounts on top.

gyre
circular surface rotation of ocean water in each major sea (a type of current). Gyres are large and permanent, and occupy the northern and southern halves of the three major oceans. Their movements are dictated by the prevailing winds and the Coriolis effect. Gyres move clockwise in the northern hemisphere and anticlockwise in the southern hemisphere.

hadal zone
deepest level of the ocean, below the abyssal zone, at depths of greater than 6,000 m/19,500 ft. The ocean trenches are in the hadal zone. There is no light in this zone and pressure is over 600 times greater than atmospheric pressure.

haematite
principal ore of iron, consisting mainly of iron(III) oxide, Fe_2O_3. It occurs as **specular haematite** (dark, metallic lustre), **kidney ore** (reddish radiating fibres terminating in smooth, rounded surfaces), and a red earthy deposit.

hanging valley
valley that joins a larger glacial trough at a higher level than the trough floor. During glaciation the ice in the smaller valley was unable to erode as deeply as the ice in the trough, and so the valley was left perched high on the side of the trough when the ice retreated. A river or stream flowing along the hanging valley often forms a waterfall as it enters the trough.

harmattan
in meteorology, a dry and dusty northeast wind that blows over West Africa.

heat island
large town or city that is warmer than the surrounding countryside. The difference in temperature is most pronounced during the winter, when the heat given off by the city's houses, offices, factories, and vehicles raises the temperature

of the air by a few degrees. The heat island effect is also caused by the presence of surfaces such as black asphalt, that absorb rather than reflect sunlight, and the lack of vegetation, which uses sunlight to photosynthesise rather than radiating it back out as heat energy.

hogback
geological formation consisting of a ridge with a sharp crest and abruptly sloping sides, the outline of which resembles the back of a hog. Hogbacks are the result of differential erosion on steeply dipping rock strata composed of alternating resistant and soft beds. Exposed, almost vertical resistant beds provide the sharp crests.

Holocene epoch
epoch of geological time that began 10,000 years ago, and continues into the present; the second and current epoch of the Quaternary period. During this epoch the glaciers retreated, the climate became warmer, and human civilizations developed significantly.

hot spot
a relatively large region of persistent volcanism that is not associated with an island arc, and not necessarily associated with an ocean spreading centre (although it can be coincident with one). Hot spots occur in the oceans and on the continents, commonly within, rather than on the edges, of lithospheric plates. Examples include Hawaii, Iceland, and Yellowstone. Hot spots are thought to be the surfice manifestations of stationary 'plumes' of hot mantle material that is continuously rising to the surface from some unknown depth in the mantle, possible the core-mantle boundary. The detailed chemistry of hot spot volcanic rocks is distinctly different from ridge or arc volcanics.

Humboldt Current
former name of the Peru Current.

hurricane tropical cyclone
severe depression (region of very low atmospheric pressure) in tropical regions, called *typhoon* in the North Pacific. It is a revolving storm originating at latitudes between 5° and 20° N or S of the Equator, when the surface temperature of the ocean is above 27°C/80°F. A central calm area, called the eye, is surrounded by inwardly spiralling winds (anticlockwise in the northern hemisphere and clockwise in the southern hemisphere) of up to 320 kph/200 mph. A hurricane is accompanied by lightning and torrential rain, and can cause extensive damage. In meteorology, a hurricane is a wind of force 12 or more on the Beaufort scale.

hydration
form of chemical weathering caused by the expansion of certain minerals as they absorb water. The expansion weakens the parent rock and may cause it to break up.

Glossary

Satellites provide invaluable information about the cloud and wind patterns of entire weather systems. This satellite picture, taken in 1969, shows a cyclonic storm, or hurricane, off Hawaii. Tropical cyclones begin in the hot, moist air over tropical oceans. As an area of very low pressure develops, air is sucked in, creating a violent storm of spiralling winds. National Aeronautical Space Agency

hydraulic action

erosive force exerted by water (as distinct from the forces exerted by rocky particles carried by water). It can wear away the banks of a river, particularly at the outer curve of a meander (bend in the river), where the current flows most strongly.

Hydraulic action occurs as a river tumbles over a waterfall to crash onto the rocks below. It will lead to the formation of a plunge pool below the waterfall. The hydraulic action of ocean waves and turbulent currents forces air into rock cracks, and therefore brings about erosion by cavitation.

hydrograph

graph showing how the discharge of a river varies with time. By studying hydrographs, water engineers can predict when flooding is likely and take action to prevent its taking place.

hydrography
study and charting of Earth's surface waters in seas, lakes, and rivers.

hydrological cycle
cycle by which water is circulated between the Earth's surface and its atmosphere. See water cycle.

hydrology
study of the location and movement of inland water, both frozen and liquid, above and below ground. It is applied to major civil engineering projects such as irrigation schemes, dams, and hydroelectric power, and in planning water supply. Hydrologic studies are also undertaken to assess drinking water supplies, to track water underground, and to understand the role of water in geological processes such as fault movement and mineral deposition.

hydrosphere
portion of the Earth made of water, ice, and water vapour, including the oceans, seas, rivers, streams, swamps, lakes, groundwater, and atmospheric water vapour. In some cases its definition is extended to include the water locked up in Earth's crust and mantle.

hydrothermal vein
crack in rock filled with minerals precipitated through the action of circulating high-temperature fluids. Igneous activity often gives rise to the circulation of heated fluids that migrate outwards and move through the surrounding rock. When such solutions carry metallic ions, ore-mineral deposition occurs in the new surroundings on cooling.

hydrothermal vent or smoker
crack in the ocean floor, commonly associated with an ocean ridge, through which hot, mineral-rich water flows into the cold ocean water, forming thick clouds of suspended material. The clouds may be dark or light, depending on the mineral content, thus producing 'white smokers' or 'black smokers'. In some cases the water is clear.

Sea water percolating through the sediments and crust is heated by the hot rocks and magma below and then dissolves minerals from the hot rocks. The water gets so hot that its increased buoyancy drives it back out into the ocean via a hydrothermal ('hot water') vent. When the water, anywhere from 60°C/140°F to over 400°C/750°F (kept liquid by the pressure of the ocean above) comes into contact with the frigid sea water, the sudden cooling causes these minerals to precipitate from solution, so forming the suspension. These minerals settle out and crystallize, forming stalagmite-like 'chimneys'. The chemical-rich water around a smoker gives rise to colonies of primitive bacteria that use the chemicals in the water, rather than the sunlight, for energy. Strange animals that live in such regions include huge tube worms 2 m/6 ft long, giant clams, and species of crab, anemone, and shrimp found nowhere else.

Glossary

hydrothermal vent
hot fissure in the ocean floor, known as a smoker.

Iapetus Ocean or Proto-Atlantic
sea that existed in early Palaeozoic times between the continent that was to become Europe and that which was to become North America. The continents moved together in the late Palaeozoic, obliterating the ocean. When they moved apart once more, they formed the Atlantic.

ice age
any period of extensive glaciation occurring in the Earth's history, but particularly that in the Pleistocene epoch, immediately preceding historic times. On the North American continent, glaciers reached as far south as the Great Lakes, and an ice sheet spread over northern Europe, leaving its remains as far south as Switzerland. There were several glacial advances separated by interglacial stages during which the ice melted and temperatures were higher than today.

Other ice ages have occurred throughout geological time: there were four in the Precambrian era, one in the Ordovician, and one at the end of the Carboniferous and beginning of the Permian. The occurrence of an ice age is governed by a combination of factors (the ***Milankovitch hypothesis***): (1) the Earth's change of attitude in relation to the Sun, that is, the way it tilts in a 41,000-year cycle and at the same time wobbles on its axis in a 22,000-year cycle, making the time of its closest approach to the Sun come at different seasons; and (2) the 92,000-year cycle of eccentricity in its orbit round the Sun, changing it from an elliptical to a near circular orbit, the severest period of an ice age coinciding with the approach to circularity. There is a possibility that the Pleistocene ice age is not yet over. It may reach another maximum in another 60,000 years.

Ice Age, Little
period of particularly severe winters that gripped northern Europe between the 13th and 17th centuries. Contemporary writings and paintings show that Alpine glaciers were much more extensive than at present, and rivers such as the Thames, which do not ice over today, were so frozen that festivals could be held on them.

iceberg
floating mass of ice, about 80% of which is submerged, rising sometimes to 100 m/300 ft above sea level. Glaciers that reach the coast become extended into a broad foot; as this enters the sea, masses break off and drift towards temperate latitudes, becoming a danger to shipping.

ice sheet
body of ice that covers a large land mass or continent; it is larger than an ice cap. During the last ice age, ice sheets spread over large parts of Europe and North America. Today there are two ice sheets, covering much of Antarctica

and Greenland. About 96% of all present-day ice is in the form of ice sheets. The ice sheet covering western Greenland has increased in thickness by 2 m/6.5 ft 1981–93; this increase is the equivalent of a 10% rise in global sea levels.

igneous rock
rock formed from the solidification of molten rock called magma. The acidic nature of this rock type means that areas with underlying igneous rock are particularly susceptible to the effects of acid rain. Igneous rocks that crystallize slowly from magma below the Earth's surface have large crystals. Examples include basalt and granite.

Igneous rocks that crystallize from magma below the Earth's surface are called *plutonic* or *intrusive*, depending on the depth of formation. They have large crystals produced by slow cooling; examples include dolerite and granite. Those extruded at the surface from lava are called *extrusive* or *volcanic*. Rapid cooling results in small crystals; basalt is an example.

impermeable rock
rock that does not allow water to pass through it – for example, clay, shale, and slate. Unlike permeable rocks, which absorb water, impermeable rocks can support rivers. They therefore experience considerable erosion (unless, like slate, they are very hard) and commonly form lowland areas.

A basalt outcrop in the French Auvergne. Basalt is a volcanic rock (type of igneous rock) which when it cools sometimes cracks along its natural planes of cleavage to produce distinctive hexagonal columns. Premaphotos Wildlife

Glossary

intertropical convergence zone (ITCZ)
area of heavy rainfall found in the tropics and formed as the trade winds converge and rise to form cloud and rain. It moves a few degrees northwards during the northern summer and a few degrees southwards during the southern summer, following the apparent movement of the Sun. The ITCZ is responsible for most of the rain that falls in Africa. The doldrums are also associated with this zone.

intrusion
mass of igneous rock that has formed by 'injection' of molten rock, or magma, into existing cracks beneath the surface of the Earth, as distinct from a volcanic rock mass which has erupted from the surface. Intrusion features include vertical cylindrical structures such as stocks, pipes, and necks; sheet structures such as dykes that cut across the strata and sills that push between them; laccoliths, which are blisters that push up the overlying rock; and batholiths, which represent chambers of solidified magma and contain vast volumes of rock.

intrusive rock or plutonic rock
igneous rock formed beneath the Earth's surface. Magma, or molten rock, cools slowly at these depths to form coarse-grained rocks, such as granite, with large crystals. (Extrusive rocks, which are formed on the surface, are generally fine-grained.) A mass of intrusive rock is called an intrusion.

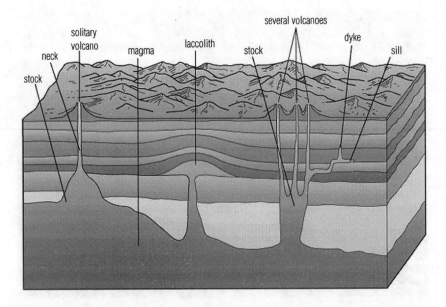

Igneous intrusions can be a variety of shapes and sizes. Laccoliths are domed circular shapes, and can be many kilometres across. Sills are intrusions that flow between rock layers. Pipes or necks connect the underlying magma chamber to surface volcanoes.

ionosphere

ionized layer of Earth's outer atmosphere (60–1,000 km/38–620 mi) that contains sufficient free electrons to modify the way in which radio waves are propagated, for instance by reflecting them back to Earth. The ionosphere is thought to be produced by absorption of the Sun's ultraviolet radiation. The British Antarctic Survey estimates that the ionosphere is decreasing at a rate of 1 km/0.6 mi every five years, based on an analysis of data 1960–98. Global warming is the probable cause.

iridium anomaly

unusually high concentrations of the element iridium found worldwide in sediments that were deposited at the Cretaceous-Tertiary boundary (K-T boundary) 65 million years ago. Since iridium is more abundant in extraterrestrial material, its presence is thought to be evidence for a large meteorite impact that may have caused the extinction of the dinosaurs and other life at the end of the Cretaceous.

island arc

curved chain of volcanic islands. Island arcs are common in the Pacific where they ring the ocean on both sides; the Aleutian Islands off Alaska are an example. The volcanism that forms island arcs is a result of subduction of an oceanic plate beneath another plate, as evidenced by the presence of ocean trenches on the convex side of the arc, and the Benioff zone of high seismic activity beneath.

Such island arcs are often later incorporated into continental margins during mountain-building episodes.

isobar

line drawn on maps and weather charts linking all places with the same atmospheric pressure (usually measured in millibars). When used in weather forecasting, the distance between the isobars is an indication of the barometric gradient (the rate of change in pressure).

Where the isobars are close together, cyclonic weather is indicated, bringing strong winds and a depression, and where far apart anticyclonic, bringing calmer, settled conditions.

isostasy

condition of gravitational equilibrium of all parts of the Earth's crust. The crust is in isostatic equilibrium if, below a certain depth, the weight and thus pressure of rocks above is the same everywhere. The idea is that the lithosphere floats on the asthenosphere as a piece of wood floats on water. A thick piece of wood floats lower than a thin piece, and a denser piece of wood floats lower than a less dense piece. There are two theories of the mechanism of isostasy, the Airy hypothesis and the Pratt hypothesis, both of which have validity. In the *Airy hypothesis* crustal blocks have the same density but different thicknesses: like ice cubes floating in water, higher mountains have deeper

Glossary

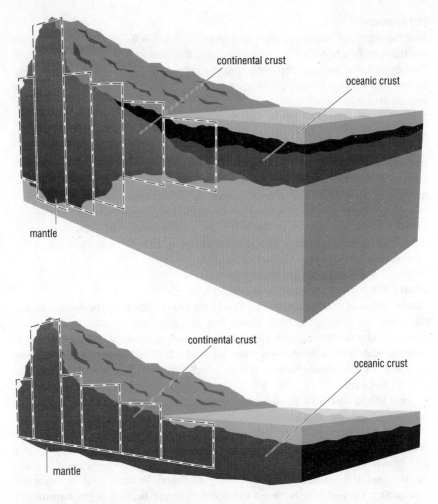

Isostasy explains the vertical distribution of Earth's crust. George Bedell Airy proposed that the density of the crust is everywhere the same and the thickness of crustal material varies. Higher mountains are compensated by deeper roots. This explains the high elevations of most major mountain chains, such as the Himalayas. G H Pratt hypothesized that the density of the crust varies, allowing the base of the crust to be the same everywhere. Sections of crust with high mountains, therefore, would be less dense than sections of crust where there are lowlands. This applies to instances where density varies, such as the difference between continental and oceanic crust.

roots. In the *Pratt hypothesis*, crustal blocks have different densities allowing the depth of crustal material to be the same. In practice, both mechanisms are at work.

isotherm
line on a map, linking all places having the same temperature at a given time.

jade

semiprecious stone consisting of either jadeite, $NaAlSi_2O_6$ (a pyroxene), or nephrite, $Ca_2(Mg,Fe)_5Si_8O_{22}(OH,F)_2$ (an amphibole), ranging from colourless through shades of green to black according to the iron content. Jade ranks 5.5–6.5 on the Mohs scale of hardness.

jet stream

narrow band of very fast wind (velocities of over 150 kph/95 mph) found at altitudes of 10–16 km/6–10 mi in the upper troposphere or lower stratosphere. Jet streams usually occur about the latitudes of the Westerlies (35°–60°).

The jet stream may be used by high flying aircraft to speed their journeys. Their discovery of the existence of the jet stream allowed the Japanese to send gas-filled balloons carrying bombs to the northwestern US during World War II.

Jurassic period

period of geological time 208–146 million years ago; the middle period of the Mesozoic era. Climates worldwide were equable, creating forests of conifers and ferns; dinosaurs were abundant, birds evolved, and limestones and iron ores were deposited.

The name comes from the Jura Mountains in France and Switzerland, where the rocks formed during this period were first studied.

kame

geological feature, usually in the form of a mound or ridge, formed by the deposition of rocky material carried by a stream of glacial meltwater. Kames are commonly laid down in front of or at the edge of a glacier (kame terrace), and are associated with the disintegration of glaciers at the end of an ice age.

Kames are made of well-sorted rocky material, usually sands and gravels. The rock particles tend to be rounded (by attrition) because they have been transported by water.

kaolinite

white or greyish clay mineral, hydrated aluminium silicate, $Al_2Si_2O_5(OH)_4$, formed mainly by the decomposition of feldspar in granite. It is made up of platelike crystals, the atoms of which are bonded together in two-dimensional sheets, between which the bonds are weak, so that they are able to slip over one another, a process made more easy by a layer of water. China clay (kaolin) is derived from it. It is mined in France, the UK, Germany, China, and the USA.

karst

landscape characterized by remarkable surface and underground forms, created as a result of the action of water on permeable limestone. The feature takes its name from the Karst region on the Adriatic coast in Slovenia and

Croatia, but the name is applied to landscapes throughout the world, the most dramatic of which is found near the city of Guilin in the Guangxi province of China.

Limestone is soluble in the weak acid of rainwater. Erosion takes place most swiftly along cracks and joints in the limestone and these open up into gullies called grikes. The rounded blocks left upstanding between them are called clints.

katabatic wind
cool wind that blows down a valley on calm clear nights. (By contrast, an anabatic wind is warm and moves up a valley in the early morning.) When the sky is clear, heat escapes rapidly from ground surfaces, and the air above the ground becomes chilled. The cold dense air moves downhill, forming a wind that tends to blow most strongly just before dawn.

Cold air blown by a katabatic wind may collect in a depression or valley bottom to create a frost hollow. Katabatic winds are most likely to occur in the late spring and autumn because of the greater daily temperature differences.

khamsin
hot southeasterly wind that blows from the Sahara desert over Egypt and parts of the Middle East from late March to May or June. It is called *sharav* in Israel.

K–T boundary
geologists' shorthand for the boundary between the rocks of the Cretaceous and the Tertiary periods 65 million years ago. It coincides with the end of the extinction of the dinosaurs and in many places is marked by a layer of clay or rock enriched in the element iridium. Extinction of the dinosaurs at the K–T boundary and deposition of the iridium layer are thought to be the result of either impact of an asteroid or comet that crashed into the Yucatán Peninsula (forming the **Chicxulub crater**), perhaps combined with a period of intense volcanism on the continent of India.

laccolith
intruded mass of igneous rock that forces apart two strata and forms a round lens-shaped mass many times wider than thick. The overlying layers are often pushed upward to form a dome. A classic development of laccoliths is illustrated in the Henry, La Sal, and Abajo mountains of southeastern Utah, USA, found on the Colorado plateau.

lahar
mudflow formed of a fluid mixture of water and volcanic ash. During a volcanic eruption, melting ice may combine with ash to form a powerful flow capable of causing great destruction. The lahars created by the eruption of Nevado del Ruiz in Colombia, South America, in 1985 buried 22,000 people in 8 m/26 ft of mud.

land reclamation

conversion of derelict or otherwise unusable areas into productive land. For example, where industrial or agricultural activities, such as sand and gravel extraction, or open-cast mining have created large areas of derelict or waste ground, the companies involved are usually required to improve the land so that it can be used.

landslide

sudden downward movement of a mass of soil or rocks from a cliff or steep slope. Landslides happen when a slope becomes unstable, usually because the base has been undercut or because materials within the mass have become wet and slippery.

A *mudflow* happens when soil or loose material is soaked so that it no longer adheres to the slope; it forms a tongue of mud that reaches downhill from a semicircular hollow. A *slump* occurs when the material stays together as a large mass, or several smaller masses, and these may form a tilted steplike structure as they slide. A *landslip* is formed when beds of rock dipping towards a cliff slide along a lower bed. Earthquakes may precipitate landslides.

land use

way in which a given area of land is used. Land is often classified according to its use, for example, for agriculture, industry, residential buildings, and recreation. The first land use surveys in the UK were conducted during the 1930s.

lapis lazuli

rock containing the blue mineral lazurite in a matrix of white calcite with small amounts of other minerals. It occurs in silica-poor igneous rocks and metamorphic limestones found in Afghanistan, Siberia, Iran, and Chile. Lapis lazuli was a valuable pigment of the Middle Ages, also used as a gemstone and in inlaying and ornamental work.

Laurasia

northern landmass formed 200 million years ago by the splitting of the single world continent Pangaea. (The southern landmass was Gondwanaland.) It consisted of what was to become North America, Greenland, Europe, and Asia, and is believed to have broken up about 100 million years ago with the separation of North America from Europe.

lava

magma that erupts from a volcano and cools to form extrusive igneous rock. Lava types differ in composition, temperature, gas content, and viscosity (resistance to flow).

The three major lava types are basalt (dark, fluid, and relativey low silica content), rhyolite (light, viscous, high silica content), and andesite (an intermediate lava).

levee
naturally formed raised bank along the side of a river channel. When a river overflows its banks, the rate of flow is less than that in the channel, and silt is deposited on the banks. With each successive flood the levee increases in size so that eventually the river may be above the surface of the surrounding flood plain. Notable levees are found on the lower reaches of the Mississippi in the USA and the Po in Italy.

limestone
sedimentary rock composed chiefly of calcium carbonate $CaCO_3$, either derived from the shells of marine organisms or precipitated from solution, mostly in the ocean. Various types of limestone are used as building stone.

limestone pavement
bare rock surface resembling a block of chocolate, found on limestone plateaus. It is formed by the weathering of limestone into individual upstanding blocks, called clints, separated from each other by joints, called grykes. The weathering process is thought to entail a combination of freeze–thaw (the alternate freezing and thawing of ice in cracks) and carbonation (the dissolving of minerals in the limestone by weakly acidic rainwater). Malham Tarn in North Yorkshire is an example of a limestone pavement.

limnology
study of lakes and other bodies of open fresh water, in terms of their plant and animal biology, chemistry, and physical properties.

lithification
conversion of an unconsolidated sediment into solid sedimentary rock by *compaction* of mineral grains that make up the sediment, *cementation* by crystallization of new minerals from percolating aqueous solutions, and new growth of the original mineral grains. The term is less commonly used to refer to solidification of magma to form igneous rock.

lithosphere
upper rocky layer of the Earth that forms the jigsaw of plates that take part in the movements of plate tectonics. The lithosphere comprises the crust and a portion of the upper mantle. It is regarded as being rigid and brittle and moves about on the more plastic and less brittle asthenosphere. The lithosphere ranges in thickness from 2–3 km/1.2–1.9 mi at mid-ocean ridges to 150 km/90 mi beneath old ocean crust, to 250 km/90 mi under cratons.

loam
type of fertile soil, a mixture of sand, silt, clay, and organic material. It is porous, which allows for good air circulation and retention of moisture.

lode
geological deposit rich in certain minerals, generally consisting of a large vein or set of veins containing ore minerals. A system of veins that can be mined directly forms a lode, for example the mother lode of the California gold rush.

loess
yellow loam, derived from glacial meltwater deposits and accumulated by wind in periglacial regions during the ice ages. Loess usually attains considerable depths, and the soil derived from it is very fertile. There are large deposits in central Europe (Hungary), China, and North America. It was first described in 1821 in the Rhine area, and takes its name from a village in Alsace.

longshore drift
movement of material along a beach. When a wave breaks obliquely, pebbles are carried up the beach in the direction of the wave (swash). The wave draws back at right angles to the beach (backwash), carrying some pebbles with it. In this way, material moves in a zigzag fashion along a beach.

Longshore drift is responsible for the erosion of beaches and the formation of spits (ridges of sand or shingle projecting into the water). Attempts are often made to halt longshore drift by erecting barriers, or groynes, at right angles to the shore.

maelstrom
whirlpool off the Lofoten Islands, Norway, also known as the Moskenesstraumen, which gave its name to whirlpools in general.

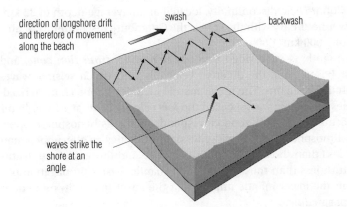

Waves sometimes hit the beach at an angle. The incoming waves (swash) carry sand and shingle up onto the shore and the outgoing wave takes some material away with it. Gradually material is carried down the shoreline in the same direction as the longshore current.

mafic rock

plutonic rock composed chiefly of dark-coloured minerals such as olivine and pyroxene that contain abundant magnesium and iron. It is derived from *magnesium* and *ferric* (iron). The term *mafic* also applies to dark-coloured minerals rich in iron and magnesium as a group. 'Mafic rocks' usually refers to dark-coloured igneous rocks such as basalt, but can also refer to their metamorphic counterparts.

magma

molten rock (either beneath or on a planetary surface) from which igneous rocks are formed. Lava is magma that has extruded on to the surface.

magnetic storm

in meteorology, a sudden disturbance affecting the Earth's magnetic field, causing anomalies in radio transmissions and magnetic compasses. It is probably caused by sunspot activity.

magnetometer

device for measuring the intensity and orientation of the magnetic field of a particular rock or of a certain area. In geology, magnetometers are used to determine the original orientation of a rock formation (or the orientation when the magnetic signature was locked in), which allows for past plate reconstruction. They are also used to delineate 'magnetic striping' on the sea floor in order to make plate reconstruction and to prospect for ore bodies such as iron ore, which can disrupt the local magnetic field.

mantle

intermediate zone of the Earth between the crust and the core, accounting for 82% of Earth's volume. The boundary between the mantle and the crust above is the Mohorovičić discontinuity, located at an average depth of 32 km/20 mi. The lower boundary with the core is the Gutenburg discontinuity at an average depth of 2,900 km/1,800 mi.

The mantle is subdivided into *upper mantle*, *transition zone*, and *lower mantle*, based upon the different velocities with which seismic waves travel through these regions. The upper mantle includes a zone characterized by low velocities of seismic waves, called the *low velocity zone*, at 72 km/45 mi to 250 km/155 mi depth. This zone corresponds to the asthenosphere upon which Earth's lithospheric plates glide. Seismic velocities in the upper mantle are overall less than those in the transition zone and those of the transition zone are in turn less than those of the lower mantle. Faster propagation of seismic waves in the lower mantle implies that the lower mantle is more dense than the upper mantle.

The mantle is composed primarily of magnesium, silicon, and oxygen in the form of silicate minerals. In the upper mantle, the silicon in silicate minerals, such as olivine, is surrounded by four oxygen atoms. Deeper in the transition zone greater pressures promote denser packing of oxygen such that some silicon is surrounded by six oxygen atoms, resulting in magnesium silicates with garnet and pyroxene structures. Deeper still, all silicon is surrounded by six oxygen atoms so that the new mineral $MgSiO_3$-perovskite predominates.

mantle keel
relatively cold slab of mantle material attached to the underside of a continental craton (core of a continent composed of old, highly deformed metamorphic rock), and protruding down into the mantle like the keel of a boat. Their presence suggests that tectonic processes may have been different at the time the cratons were formed.

marble
rock formed by metamorphosis of sedimentary limestone. It takes and retains a good polish, and is used in building and sculpture. In its pure form it is white and consists almost entirely of calcite $CaCO_3$. Mineral impurities give it various colours and patterns. Carrara, Italy, is known for white marble.

Mariana Trench
lowest region on the Earth's surface; the deepest part of the sea floor. The trench is 2,400 km/1,500 mi long and is situated 300 km/200 mi east of the Mariana Islands, in the northwestern Pacific Ocean. Its deepest part is the gorge known as the Challenger Deep, which extends 11,034 m/36,210 ft below sea level.

mass extinction
event that produces the extinction of many species at about the same time. One notable example is the boundary between the Cretaceous and Tertiary periods (known as the K-T boundary) that saw the extinction of the dinosaurs and other large reptiles, and many of the marine invertebrates as well. Mass extinctions have taken place frequently during Earth's history.

There have been five major mass extinctions, in which 75% or more of the world's species have been wiped out: End Ordovician period (440 million years ago) in which about 85% of species were destroyed (second most severe); Late Devonian period (365 million years ago) which took place in two waves a million years apart, and was the third most severe, with marine species particularly badly hit; Late Permian period (251 million years ago), the gravest mass extinction in which about 96% of species became extinct; Late Triassic (205 million years ago), in which about 76% of species were destroyed, mainly marine; Late Cretaceous period (65 million years ago), in which 75–80% of species became extinct, including dinosaurs.

meander
loop-shaped curve in a mature river flowing sinuously across flat country. As a river flows, any curve in its course is accentuated by the current. On the outside of the curve the velocity, and therefore the erosion, of the current is greatest. Here the river cuts into the outside bank, producing a **cutbank** or **river cliff** and the river's deepest point, or **thalweg**. On the curve's inside the current

The course of a river from its source of a spring or melting glacier, through to maturity where it flows into the sea.

is slow and deposits any transported material, building up a gentle slip-off slope. As each meander migrates in the direction of its cutbank, the river gradually changes its course across the flood plain.

A loop in a river's flow may become so accentuated that it becomes cut off from the normal course and forms an oxbow lake. The occurrence of meanders depends upon the gradient (slope) of the land, the nature of the river's discharge, and the type of material being carried. Meanders are common where the gradient is gentle, the discharge fairly steady (not subject to extremes), and the material carried is fine. The word meander comes from the River Menderes in Turkey.

mechanical weathering
alternative name for physical weathering.

medial moraine
linear ridge of rocky debris running along the centre of a glacier. Medial moraines are commonly formed by the joining of two lateral moraines when two glaciers merge.

Mercalli scale
qualitative scale of the intensity of an earthquake. It differs from the Richter scale, which indicates earthquake *magnitude* and is quantitative. It is named after the Italian seismologist Giuseppe Mercalli (1850–1914).

Intensity is a subjective value, based on observed phenomena, and varies from place to place with the same earthquake (as opposed to the Richter scale, which is absolute).

mesa
flat-topped, steep-sided plateau, consisting of horizontal weak layers of rock topped by a resistant formation; in particular, those found in the desert areas of the USA and Mexico. A small mesa is called a butte.

mesosphere
layer in the Earth's atmosphere above the stratosphere and below the thermosphere. It lies between about 50 km/31 mi and 80 km/50 mi above the ground.

Mesozoic
era of geological time 245–65 million years ago, consisting of the Triassic, Jurassic, and Cretaceous periods. At the beginning of the era, the continents were joined together as Pangaea; dinosaurs and other giant reptiles dominated the sea and air; and ferns, horsetails, and cycads thrived in a warm climate worldwide. By the end of the Mesozoic era, the continents had begun to assume their present positions, flowering plants were dominant, and many of the large reptiles and marine fauna were becoming extinct.

metamorphic rock
rock that has been changed from its original form, texture, and/or mineral assemblage by pressure or heat. For example, limestone can be metamorphosed by heat into marble; shale by pressure into slate. The term was coined in 1833 by Scottish geologist Charles Lyell.

There are two main types of metamorphism. *Thermal metamorphism*, or contact metamorphism, is brought about by the baking of solid rocks in the vicinity of an igneous intrusion (molten rock, or magma, in a crack in the Earth's crust). It is responsible, for example, for the conversion of limestone to marble. *Regional metamorphism* results from the heat and intense pressures associated with burial and the movements and collision of tectonic plates (see plate tectonics). It brings about the conversion of shale to slate, for example. A third type, *shock metamorphism*, occurs when a rock is very quickly subjected to high pressures such as those brought about by a meteorite impact.

Metamorphic rocks have essentially the same chemical composition as their protoliths, but because different minerals are stable at different temperatures and pressures, they are commonly composed of different, metamorphic minerals. Metamorphism also produces changes in the texture of the rock, for example, rocks become foliated (or layered), and crystals grow larger.

metamorphism

geological term referring to the changes in rocks of the Earth's crust caused by increasing pressure and temperature. The resulting rocks are metamorphic rocks. All metamorphic changes take place in solid rocks. If the rocks melt and then harden, they are considered igneous rocks.

meteorite

piece of rock or metal from space that reaches the surface of the Earth, Moon, or other body. Most meteorites are thought to be fragments from asteroids, although some may be pieces from the heads of comets. Most are stony, although some are made of iron and a few have a mixed rock-iron composition.

Stony meteorites can be divided into two kinds: ***chondrites*** and ***achondrites***. Chondrites contain chondrules, small spheres of the silicate minerals olivine and orthopyroxene, and comprise 85% of meteorites. Achondrites do not contain chondrules. Meteorites provide evidence for the nature of the Solar System and may be similar to the Earth's core and mantle, neither of which can be observed directly.

Thousands of meteorites hit the Earth each year, but most fall in the sea or in remote areas and are never recovered. The largest known meteorite is one composed of iron, weighing 60 tonnes, which lies where it fell in prehistoric times at Grootfontein, Namibia. Meteorites are slowed down by the Earth's atmosphere, but if they are moving fast enough they can form a crater on impact. Meteor Crater in Arizona, about 1.2 km/0.7 mi in diameter and 200 m/650 ft deep, is the site of a meteorite impact about 50,000 years ago.

meteorology

scientific observation and study of the atmosphere, so that weather can be accurately forecast.

Data from meteorological stations and weather satellites are collated by computer at central agencies, and forecasts and weather maps based on current readings are issued at regular intervals. Modern analysis, employing some of the most powerful computers, can give useful forecasts for up to six days ahead.

At meteorological stations readings are taken of the factors determining weather conditions: atmospheric pressure, temperature, humidity, wind (using the Beaufort scale), cloud cover (measuring both type of cloud and coverage), and precipitation such as rain, snow, and hail (measured at 12-hour intervals). Satellites are used either to relay information transmitted from the Earth-based stations, or to send pictures of cloud development, indicating wind patterns, and snow and ice cover.

History
Apart from some observations included by Aristotle in his book *Meteorologia*, meteorology did not become a precise science until the end of the 16th century, when Galileo and the Florentine academicians constructed the first thermometer of any importance, and when Evangelista Torricelli in 1643 discovered the principle of the barometer. Robert Boyle's work on gases, and that of his assistant, Robert Hooke, on barometers, advanced the physics necessary for the understanding of the weather. Gabriel Fahrenheit's invention of a superior mercury thermometer provided further means for temperature recording.

Weather maps
In the early 19th century a chain of meteorological stations was established in France, and weather maps were constructed from the data collected. The first weather map in England, showing the trade winds and monsoons, was made in 1688, and the first telegraphic weather report appeared 31 August 1848. The first daily telegraphic weather map was prepared at the Great Exhibition in 1851, but the Meteorological Office was not established in London until 1855. The first regular daily collections of weather observations by telegraph and the first British daily weather reports were made in 1860, and the first daily printed maps appeared 1868.

Collecting data
Observations can be collected not only from land stations, but also from weather ships, aircraft, and self-recording and automatic transmitting stations, such as the radiosonde. Radar may be used to map clouds and storms. Satellites have played an important role in televising pictures of global cloud distribution.

As well as supplying reports for the media, the Meteorological Office in Bracknell, near London, does specialist work for industry, agriculture, and transport. Kew is the main meteorological observatory in the British Isles, but other observatories are at Eskdalemuir in the southern uplands of Scotland, Lerwick in the Shetlands, and Valentia in southwestern Ireland. Climatic information from British climatological reporting stations is published in the Monthly Weather Report, and periodically in tables of averages and frequencies.

The British Meteorological Office's Daily Weather Report contains a detailed map of the weather over the British Isles and a less detailed map of the weather over the northern hemisphere, and the Daily Aerological Record contains full reports of radiosonde ascents made over the British Isles and from some of the ocean weather ships, together with maps of the heights of the 700 millibars (mb), 500 mb, and 300 mb pressure surfaces, giving a picture of the winds at 3,048 m/10,000 ft, 5,182 m/18,000 ft, and 9,144 m/30,000 ft; there is also a map of the height of the tropopause. Ships' reports are plotted on the same charts using the same symbolic form. Data from radiosondes and aircraft are plotted on upper-air charts and on temperature–height diagrams, the diagram in use in Britain being the tephigram. With the help of this diagram it is possible to predict the formation or otherwise of clouds, showers, or thunderstorms, and sometimes to identify the source region of the air mass.

microclimate

climate of a small area, such as a woodland, lake, or even a hedgerow. Significant differences can exist between the climates of two neighbouring areas – for example, a town is usually warmer than the surrounding countryside (forming a heat island), and a woodland cooler, darker, and less windy than an area of open land.

Microclimates play a significant role in agriculture and horticulture, as different crops require different growing conditions.

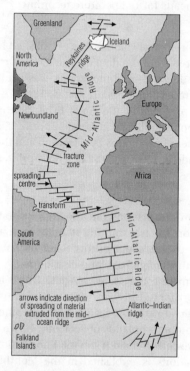

The Mid-Atlantic Ridge is the boundary between the crustal plates that form America, and Europe and Africa. An oceanic ridge cannot be curved since the material welling up to form the ridge flows at a right angle to the ridge. The ridge takes the shape of small straight sections offset by fractures transverse to the main ridge.

Mid-Atlantic Ridge

ocean ridge that runs along the centre of the Atlantic Ocean, parallel to its edges, for some 14,000 km/ 8,800 mi – almost from the Arctic to the Antarctic. Like other ocean ridges, the Mid-Atlantic Ridge is essentially a linear, segmented volcano.

The Mid-Atlantic Ridge is central because the ocean crust beneath the Atlantic Ocean has continually grown outwards from the ridge at a steady rate during the past 200 million years. Iceland straddles the ridge and was formed by volcanic outpourings.

mid-ocean ridge

long submarine mountain range that winds along the middle of the ocean floor for roughly 60,000 km/ 37,300 mi. The mid-ocean ridge system is essentially a segmented, linear shield volcano. There are a number of major ridges, including the Mid-Atlantic Ridge, which runs down the centre of the Atlantic; the East Pacific Rise in the southeast Pacific; and the Southeast Indian Ridge. Ridges are now known to be spreading centres, or divergent margins where two plates of oceanic lithosphere are moving away from one another (see plate tectonics). Ocean ridges can rise thousands of metres above the surrounding seabed.

Ocean ridges usually have a rift valley along their crests, indicating where the flanks are being pulled apart by the growth of the plates of the lithosphere beneath. The crests are generally free of sediment; increasing depths of sediment are found with increasing distance down the flanks.

Milankovitch hypothesis

combination of factors governing the occurrence of ice ages proposed in 1930 by the Yugoslav geophysicist M Milankovitch (1879–1958). These include the variation in the angle of the Earth's axis, and the geometry of the Earth's orbit around the Sun.

mineral

naturally formed inorganic substance with a particular chemical composition and a regularly repeating internal structure. Either in their perfect crystalline form or otherwise, minerals are the constituents of rocks. In more general usage, a mineral is any substance economically valuable for mining (including coal and oil, despite their organic origins).

Mineral forming processes include: melting of pre-existing rock and subsequent crystallization of a mineral to form magmatic or volcanic rocks; weathering of rocks exposed at the land surface, with subsequent transport and grading by surface waters, ice or wind to form sediments; and recrystallization through increasing temperature and pressure with depth to form metamorphic rocks.

Minerals are usually classified as magmatic, sedimentary, or metamorphic. The magmatic minerals include the feldspars, quartz, pyroxenes, amphiboles, micas, and olivines that crystallize from silica-rich rock melts within the crust or from extruded lavas.

The most commonly occurring sedimentary minerals are either pure concentrates or mixtures of sand, clay minerals, and carbonates (chiefly calcite, aragonite, and dolomite).

Minerals typical of metamorphism include andalusite, cordierite, garnet, tremolite, lawsonite, pumpellyite, glaucophane, wollastonite, chlorite, micas, hornblende, staurolite, kyanite, and diopside.

mineralogy

study of minerals. The classification of minerals is based chiefly on their chemical composition and the kind of chemical bonding that holds these atoms together. The mineralogist also studies their crystallographic and physical characters, occurrence, and mode of formation.

The systematic study of minerals began in the 18th century, with the division of minerals into four classes: earths, metals, salts, and bituminous substances, distinguished by their reactions to heat and water.

Miocene

fourth epoch of the Tertiary period of geological time, 23.5–5.2 million years ago. At this time grasslands spread over the interior of continents, and hoofed mammals rapidly evolved.

Mississippian
US term for the Lower or Early Carboniferous period of geological time, 363–323 million years ago. It is named after the state of Mississippi.

mistral
cold, dry, northerly wind that occasionally blows during the winter on the Mediterranean coast of France, particularly concentrated along the Rhône valley. It has been known to reach a velocity of 145 kph/90 mph.

model
simplified version of some aspect of the real world. Models are produced to show the relationships between two or more factors, such as land use and the distance from the centre of a town (for example, concentric-ring theory). Because models are idealized, they give only a general guide to what may happen.

Mohole
US project for drilling a hole through the Earth's crust, so named from the Mohorovičić discontinuity that marks the transition from crust to mantle. Initial tests were made in the Pacific in 1961, but the project was subsequently abandoned.

The cores that were brought up illuminated the geological history of the Earth and aided the development of geophysics.

Mohorovičić discontinuity also Moho or M-discontinuity
seismic discontinuity, marked by a rapid increase in the speed of earthquake waves, that is taken to represent the boundary between the Earth's crust and mantle. It follows the variations in the thickness of the crust and is found approximately 35–40 km/22–25 mi below the continents and about 10 km/6 mi below the oceans. It is named after the Croatian geophysicist Andrija Mohorovičić, who suspected its presence after analysing seismic waves from the Kulpa Valley earthquake in 1909. The 'Moho' is as deep as 70 km/143 mi beneath high mountain ranges.

Mohs scale
scale of hardness for minerals (in ascending order): 1 talc; 2 gypsum; 3 calcite; 4 fluorite; 5 apatite; 6 orthoclase; 7 quartz; 8 topaz; 9 corundum; 10 diamond.

monsoon
wind pattern that brings seasonally heavy rain to South Asia; it blows towards the sea in winter and towards the land in summer. The monsoon may cause destructive flooding all over India and Southeast Asia from April to September, leaving thousands of people homeless each year.

The monsoon cycle is believed to have started about 12 million years ago with the uplift of the Himalayas.

moraine

rocky debris or till carried along and deposited by a glacier. Material eroded from the side of a glaciated valley and carried along the glacier's edge is called a *lateral moraine*; that worn from the valley floor and carried along the base of the glacier is called a *ground moraine*. Rubble dropped at the snout of a melting glacier is called a **terminal moraine**.

When two glaciers converge their lateral moraines unite to form a *medial moraine*. Debris that has fallen down crevasses and becomes embedded in the ice is termed an *englacial moraine*; when this is exposed at the surface due to partial melting it becomes ablation moraine.

mountain

natural upward projection of the Earth's surface, higher and steeper than a hill. Mountains are at least 330 m/1,000 ft above the surrounding topography. The existing rock is also subjected to high temperatures and pressures causing metamorphism. Plutonic activity also can accompany mountain building.

natural hazard

naturally occurring phenomenon capable of causing destruction, injury, disease, or death. Examples include earthquakes, floods, hurricanes, or famine. Natural hazards occur globally and can play an important role in shaping the landscape. The events only become hazards where people are affected. Because of this, natural hazards are usually measured in terms of the damage they cause to persons or property. Human activities can trigger natural hazards, for example skiers crossing the top of a snowpack may cause an avalanche.

network

system of nodes (junctions) and links (transport routes) through which goods, services, people, money, or information flow. Networks are often shown on topological maps.

New Madrid seismic fault zone

largest system of geological faults in the eastern USA, centred on New Madrid, Missouri. Geologists estimate that there is a 90% chance of a magnitude 6 earthquake in the area by the year 2000. This would cause much damage because the solid continental rocks would transmit the vibrations over a wide area, and buildings in the region have not been designed with earthquakes in mind.

nitre or saltpetre

potassium nitrate, KNO_3, a mineral found on and just under the ground in desert regions; used in explosives. Nitre occurs in Bihar, India, Iran, and Cape Province, South Africa. The salt was formerly used for the manufacture of gunpowder, but the supply of nitre for explosives is today largely met by making the salt from nitratine (also called Chile saltpetre, $NaNO_3$). Saltpetre is a preservative and is widely used for curing meats.

Glossary

noctilucent cloud
clouds of ice forming in the upper atmosphere at around 83 km/52 mi. They are visible on summer nights, particularly when sunspot activity is low.

node
in oceanography, the point on a stationary wave (an oscillating wave with no progressive motion) at which vertical motion is least and horizontal motion is greatest. In earth science, it is the point on a fault where the apparent motion has changed direction.

North Atlantic Drift
warm ocean current in the North Atlantic Ocean; an extension of the Gulf Stream. It flows east across the Atlantic and has a mellowing effect on the climate of northwestern Europe, particularly the British Isles and Scandinavia.

nuée ardente
rapidly flowing, glowing white-hot cloud of ash and gas emitted by a volcano during a violent eruption. The ash and other pyroclastics in the lower part of the cloud behave like an ash flow. In 1902 a nuée ardente produced by the eruption of Mount Pelee in Martinique swept down the volcano in a matter of seconds and killed 28,000 people in the nearby town of St Pierre.

nutrient cycle
transfer of nutrients from one part of an ecosystem to another. Trees, for example, take up nutrients such as calcium and potassium from the soil through their root systems and store them in leaves. When the leaves fall they are decomposed by bacteria and the nutrients are released back into the soil where they become available for root uptake again.

obsidian
black or dark-coloured glassy volcanic rock, chemically similar to granite, but formed by cooling rapidly on the Earth's surface at low pressure.

obsidian hydration-rim dating
method of dating artefacts made from the volcanic glass obsidian. Water molecules absorbed by inward diffusion through cut surfaces cause the outer areas of an obsidian article to convert to the mineral perlite. An object may be dated by measuring the thickness of this perlite – the *hydration* (combined with water) rim.

Only a molecule-thick water film is required at the surface to maintain the process, an amount available even in the near-arid zones of Egypt. Temperature, sunlight, and different chemical compositions cause variation in the hydration rate; therefore the method needs to be calibrated against an established chronological sequence for absolute dating. The method has been applied to many periods, including the Aztec age in Mexico, the pre-ceramic era of Japan (about

23,000 BC), and the tribal-war periods of Easter Island before the arrival of traders and missionaries 1722. Obsidian hydration-rim dating was stimulated by the early research of Irving Friedman of the US Geological Survey from 1955.

occluded front
weather front formed when a cold front catches up with a warm front. It brings clouds and rain as air is forced to rise upwards along the front, cooling and condensing as it does so.

ocean
great mass of salt water. Geographically speaking three oceans exist – the Atlantic, Indian, and Pacific – to which the Arctic is often added; but they are often considered a single entity. They cover approximately 70% or 363 million sq km/140 million sq mi of the total surface area of the Earth. Water levels recorded in the world's oceans have shown an increase of 10–15 cm/4–6 in over the past 100 years.

depth (average) 3,660 m/12,000 ft, but shallow ledges (continental shelves) 180 m/600 ft run out from the continents, beyond which the continental slope reaches down to the abyssal zone, the largest area, ranging from 2,000–6,000 m/6,500–19,500 ft. Only the deep-sea trenches go deeper, the deepest recorded being 11,034 m/36,201 ft (by the *Vityaz*, USSR) in the Mariana Trench of the western Pacific in 1957

features deep trenches (off eastern and southeast Asia, and western South America), volcanic belts (in the western Pacific and eastern Indian Ocean), and ocean ridges (in the mid-Atlantic, eastern Pacific, and Indian Ocean)

temperature varies on the surface with latitude (−2°C to +29°C); decreases rapidly to 370 m/1,200 ft, then more slowly to 2,200 m/7,200 ft; and hardly at all beyond that

water contents salinity averages about 3%; minerals commercially extracted include bromine, magnesium, potassium, salt; those potentially recoverable include aluminium, calcium, copper, gold, manganese, silver

pollution Oceans have always been used as a dumping area for human waste, but as the quantity of waste increases, and land areas for dumping it diminish, the problem is exacerbated. Today ocean pollutants include airborne emissions from land (33% by weight of total marine pollution); oil from both shipping and land-based sources; toxins from industrial, agricultural, and domestic uses; sewage; sediments from mining, forestry, and farming; plastic litter; and radioactive isotopes. Thermal pollution by cooling water from power plants or other industry is also a problem, killing coral and other temperature-sensitive sedentary species.

ocean current
fast-flowing body of seawater forced by the wind or by variations in water density (as a result of temperature or salinity variations) between two areas. Ocean currents are partly responsible for transferring heat from the Equator to the poles and thereby evening out the global heat imbalance.

Ocean Drilling Program (ODP, formerly the Deep-Sea Drilling Project 1968–85)

research project initiated in the USA to sample the rocks of the ocean crust. Initially under the direction of Scripps Institution of Oceanography, the project was planned and administered by the Joint Oceanographic Institutions for Deep Earth Sampling (JOIDES). The operation became international in 1975, when Britain, France, West Germany, Japan, and the USSR also became involved.

Boreholes were drilled in all the oceans using the JOIDES ships *Glomar Challenger* and *Resolution*. Knowledge of the nature and history of the ocean basins was increased dramatically. The technical difficulty of drilling the seabed to a depth of 2,000 m/6,500 ft was overcome by keeping the ship in position with side-thrusting propellers and satellite navigation, and by guiding the drill using a radiolocation system. The project is intended to continue until 2005.

oceanography

study of the oceans. Its subdivisions deal with each ocean's extent and depth, the water's evolution and composition, its physics and chemistry, the bottom topography, currents and wind, tidal ranges, biology, and the various aspects of human use. Computer simulations are widely used in oceanography to plot the possible movements of the waters, and many studies are carried out by remote sensing.

ocean trench

arcuate, submarine valley. Ocean trenches are characterized by the presense of a volcanic arc on the concave side of the trench. Trenches are now known to be related to subduction zones, places where a plate of oceanic lithosphere dives beneath another plate of either oceanic or continental lithosphere. Ocean trenches are found around the edge of the Pacific Ocean and the northeastern Indian Ocean; minor ones occur in the Caribbean and near the Falkland Islands.

Ocean trenches represent the deepest parts of the ocean floor, the deepest being the Mariana Trench which has a depth of 11,034 m/36,201 ft. At depths of below 6 km/3.6 mi there is no light and very high pressure; ocean trenches are inhabited by crustaceans, coelenterates (for example, sea anemones), polychaetes (a type of worm), molluscs, and echinoderms.

Oligocene epoch

third epoch of the Tertiary period of geological time, 35.5–3.25 million years ago. The name, from Greek, means 'a little recent', referring to the presence of the remains of some modern types of animals existing at that time.

onyx

semiprecious variety of chalcedonic silica (SiO_2) in which the crystals are too fine to be detected under a microscope, a state known as cryptocrystalline. It has straight parallel bands of different colours: milk-white, black, and red.

opal
form of hydrous silica ($SiO_2 \cdot nH_2O$), often occurring as stalactites and found in many types of rock. The common opal is translucent, milk-white, yellow, red, blue, or green, and lustrous. Precious opal is opalescent, the characteristic play of colours being caused by close-packed silica spheres diffracting light rays within the stone.

Ordnance Survey (OS)
official body responsible for the mapping of Britain. It was established in 1791 as the *Trigonometrical Survey* to continue work initiated in 1784 by Scottish military surveyor General William Roy (1726–1790). Its first accurate maps appeared in 1830, drawn to a scale of 1 in to the mile (1:63,000). In 1858 the OS settled on a scale of 1:2,500 for the mapping of Great Britain and Ireland (higher for urban areas, lower for uncultivated areas).

Subsequent revisions and editions include the 1:50,000 Landranger series of 1971–86. In 1989, the OS began using a computerized system for the creation and continuous revision of maps. Customers can now have maps drafted to their own specifications, choosing from over 50 features (such as houses, roads, and vegetation). Since 1988 the OS has had a target imposed by the government to recover all its costs from sales.

Ordovician period
period of geological time 510–439 million years ago; the second period of the Palaeozoic era. Animal life was confined to the sea: reef-building algae and the first jawless fish are characteristic.

The period is named after the Ordovices, an ancient Welsh people, because the system of rocks formed in the Ordovician period was first studied in Wales.

ore
body of rock, a vein within it, or a deposit of sediment, worth mining for the economically valuable mineral it contains. The term is usually applied to sources of metals. Occasionally metals are found uncombined (native metals), but more often they occur as compounds such as carbonates, sulphides, or oxides. The ores often contain unwanted impurities that must be removed when the metal is extracted.

orogenesis
in its original, literal sense, orogenesis means 'mountain building', but today it more specifically refers to the tectonics of mountain building (as opposed to mountain building by erosion).

Orogenesis is brought about by the movements of the rigid plates making up the Earth's crust and upper-most mantle (described by plate tectonics). Where two plates collide at a destructive margin, rocks become folded and lifted to form chains of mountains (such as the Himalayas). Processes associated with orogeny are faulting and thrusting (see fault), folding, metamorphism, and

Glossary

plutonism (see plutonic rock). However, many topographical features of mountains – cirques, u-shaped valleys – are the result of *non-orogenic* processes, such as weathering, erosion, and glaciation. Isostasy (uplift due to the buoyancy of the Earth's crust) can also influence mountain physiography.

orographic rainfall
rainfall that occurs when an airmass is forced to rise over a mountain range. As the air rises, it cools. The amount of moisture that air can hold decreases with decreasing temperature. So the water vapour in the rising airstream condenses, and rain falls on the windward side of the mountain. The air descending on the leeward side contains less moisture, resulting in a *rainshadow* where there is little or no rain.

outwash
sands and gravels deposited by streams of meltwater (water produced by the melting of a glacier). Such material may be laid down ahead of the glacier's snout to form a large flat expanse called an *outwash plain*.

Outwash is usually well sorted, the particles being deposited by the meltwater according to their size – the largest are deposited next to the snout while finer particles are deposited further downstream.

oxbow lake
curved lake found on the flood plain of a river. Oxbows are caused by the loops of meanders being cut off at times of flood and the river subsequently adopting a shorter course. In the USA, the term bayou is often used.

oxidation
form of chemical weathering caused by the chemical reaction that takes place between certain iron-rich minerals in rock and the oxygen in water. It tends to result in the formation of a red-coloured soil or deposit. The inside walls of canal tunnels and bridges often have deposits formed in this way.

Palaeocene epoch
first epoch of the Tertiary period of geological time, 65–56.5 million years ago. Many types of mammals spread rapidly after the disappearance of the great reptiles of the Mesozoic. Flying mammals replaced the flying reptiles, swimming mammals replaced the swimming reptiles, and all the ecological niches vacated by the reptiles were adopted by mammals.

At the end of the Palaeocene there was a mass extinction that caused more than half of all bottom-dwelling organisms to disappear worldwide, over a period of around one thousand years. Surface-dwelling organisms remained unaffected, as did those on land. The cause of this extinction remains unknown, though US palaeontologists have found evidence (released 1998) that it may have been caused by the Earth releasing tonnes of methane into the oceans causing increased water temperatures.

palaeomagnetic stratigraphy
use of distinctive sequences of magnetic polarity reversals to date rocks. Magnetism retained in rocks at the time of their formation are matched with known dated sequences of polar reversals or with known patterns of secular variation.

palaeomagnetism
study of the magnetic properties of rocks in order to reconstruct the Earth's ancient magnetic field and the former positions of the continents, using traces left by the Earth's magnetic field in igneous rocks before they cool. Palaeomagnetism shows that the Earth's magnetic field has reversed itself – the magnetic north pole becoming the magnetic south pole, and vice versa – at approximate half-million-year intervals, with shorter reversal periods in between the major spans.

palaeontology
study of ancient life, encompassing the structure of ancient organisms and their environment, evolution, and ecology, as revealed by their fossils and the rocks those fossils are found in. The practical aspects of palaeontology are based on using the presence of different fossils to date particular rock strata and to identify rocks that were laid down under particular conditions; for instance, giving rise to the formation of oil.

The use of fossils to trace the age of rocks was pioneered in Germany by Johann Friedrich Blumenbach (1752–1830) at Göttingen, followed by Georges Cuvier and Alexandre Brongniart in France in 1811.

Palaeozoic era
era of geological time 570–245 million years ago. It comprises the Cambrian, Ordovician, Silurian, Devonian, Carboniferous, and Permian periods. The Cambrian, Ordovician, and Silurian constitute the Lower or Early Palaeozoic; the Devonian, Carboniferous, and Permian make up the Upper or Late Palaeozoic. The era includes the evolution of hard-shelled multicellular life forms in the sea; the invasion of land by plants and animals; and the evolution of fish, amphibians, and early reptiles.

Pangaea or Pangea
single land mass, made up of all the present continents, believed to have existed between 300 and 200 million years ago; the rest of the Earth was covered by the Panthalassa ocean. Pangaea split into two land masses – Laurasia in the north and Gondwanaland in the south – which subsequently broke up into several continents. These then moved slowly to their present positions (see plate tectonics).

The former existence of a single 'supercontinent' was proposed by German meteorologist Alfred Wegener in 1912.

Glossary

Panthalassa
ocean that covered the surface of the Earth not occupied by the world continent Pangaea between 300 and 200 million years ago.

passive margin
in plate tectonics, a boundary between oceanic and continental crust that is not tectonically active. A passive boundary exists within a single tectonic plate rather than between two plates (for example the Atlantic margin of North America).

pelagic
of or pertaining to the open ocean, as opposed to bottom or shore areas.
Pelagic sediment is fine-grained fragmental material that has settled from the surface waters, usually the siliceous and calcareous skeletal remains of marine organisms, such as radiolarians and foraminifera.

Pennsylvanian period
US term for the Upper or Late Carboniferous period of geological time, 323–290 million years ago; it is named after the US state, which contains vast coral deposits.

periglacial
bordering a glacial area but not actually covered by ice, or having similar climatic and environmental characterisitics, such as mountainous areas. Periglacial areas today include parts of Siberia, Greenland, and North America. The rock and soil in these areas is frozen to a depth of several metres (permafrost) with only the top few centimetres thawing during the brief summer. The vegetation is characteristic of tundra.

During the last ice age all of southern England was periglacial. Weathering by freeze–thaw (the alternate freezing and thawing of ice in rock cracks) would have been severe, and solifluction would have taken place on a large scale, causing wet topsoil to slip from valley sides.

permafrost
condition in which a deep layer of soil does not thaw out during the summer. Permafrost occurs under periglacial conditions. It is claimed that 26% of the world's land surface is permafrost.

Permafrost gives rise to a poorly drained form of grassland typical of northern Canada, Siberia, and Alaska known as tundra.

permeable rock
rock which through which water can pass either via a network of spaces between particles or along bedding planes, cracks, and fissures. Permeable rocks can become saturated. Examples of permeable rocks include limestone (which is heavily jointed) and chalk (porous).

Unlike impermeable rocks, which do not allow water to pass through, permeable rocks rarely support rivers and are therefore subject to less erosion. As a result they commonly form upland areas (such as the chalk downs of southeastern England, and the limestone Pennines of northern England).

Permian
period of geological time 290–245 million years ago, the last period of the Palaeozoic era. Its end was marked by a dramatic change in marine life – the greatest mass extinction in geological history – including the extinction of many corals and trilobites. Deserts were widespread, terrestrial amphibians and mammal-like reptiles flourished, and cone-bearing plants (gymnosperms) came to prominence. In the oceans, 49% of families and 72% of genera vanished in the late Permian. On land, 78% of reptile families and 67% of amphibian families disappeared.

Peru Current formerly known as the Humboldt Current
cold ocean current flowing north from the Antarctic along the west coast of South America to southern Ecuador, then west. It reduces the coastal temperature, making the western slopes of the Andes arid because winds are already chilled and dry when they meet the coast.

petroleum or crude oil
natural mineral oil, a thick greenish-brown flammable liquid found underground in permeable rocks. Petroleum consists of hydrocarbons mixed with oxygen, sulphur, nitrogen, and other elements in varying proportions. It is thought to be derived from ancient organic material that has been converted by, first, bacterial action, then heat, and pressure (but its origin may be chemical also).

petrology
branch of geology that deals with the study of rocks, their mineral compositions, their textures, and their origins.

Phanerozoic eon
eon in Earth history, consisting of the most recent 570 million years. It comprises the Palaeozoic, Mesozoic, and Cenozoic eras. The vast majority of fossils come from this eon, owing to the evolution of hard shells and internal skeletons. The name means 'interval of well-displayed life'.

pingo
landscape feature of tundra terrain consisting of a hemispherical mound about 30 m/100 ft high, covered with soil that is cracked at the top. The core consists of ice, probably formed from the water of a former lake. The lake that forms when such a feature melts after an ice age is also called a pingo.

plate or tectonic plate, or lithospheric plate

one of several relatively distinct sections of lithosphere approximately 100 km/60 mi thick, which together comprise the outermost layer of the Earth like the pieces of the cracked shell of a hard-boiled egg.

The plates are made up of two types of crustal material: oceanic crust (sima) and continental crust (sial), both of which are underlain by a solid layer of mantle. Dense *oceanic crust* lies beneath Earth's oceans and consists largely of basalt. *Continental crust*, which underlies the continents and their continental shelves, is thicker, less dense, and consists of rocks rich in silica and aluminium.

Due to convection in the Earth's mantle (see plate tectonics) these pieces of lithosphere are in motion, riding on a more plastic layer of the mantle, called the asthenosphere. Mountains, volcanoes, earthquakes, and other geological features and phenomena all come about as a result of interaction between the plates.

plateau

elevated area of fairly flat land, or a mountainous region in which the peaks are at the same height. An *intermontane plateau* is one surrounded by mountains. A *piedmont plateau* is one that lies between the mountains and low-lying land. A *continental plateau* rises abruptly from low-lying lands or the sea. Examples are the Tibetan Plateau and the Massif Central in France.

plate tectonics

theory formulated in the 1960s to explain the phenomena of continental drift and seafloor spreading, and the formation of the major physical features of the Earth's surface. The Earth's outermost layer, the lithosphere, is regarded as a jigsaw puzzle of rigid major and minor plates that move relative to each other, probably under the influence of convection currents in the mantle beneath. At the margins of the plates, where they collide or move apart or slide past one another, major landforms such as mountains, rift valleys, volcanoes, ocean trenches, and *ocean ridges* are created. The rate of plate movement is at most 15 cm/6 in per year.

The concept of plate tectonics brings together under one unifying theory many phenomena observed in the Earth's crust that were previously thought to be unrelated. The size of the crust plates is variable, as they are constantly changing, but six or seven large plates now cover much of the Earth's surface, the remainder being occupied by a number of smaller plates. Each large plate may include both continental and ocean lithosphere. As a result of seismic studies it is known that the lithosphere is a rigid layer extending to depths of 50–100 km/30–60 mi, overlying the upper part of the mantle (the asthenosphere), which is composed of rocks very close to melting point, with a low shear strength. This zone of mechanical weakness allows the movement of the overlying plates. The margins of the plates are defined by major earthquake zones and belts of volcanic and tectonic activity, which have been well known for

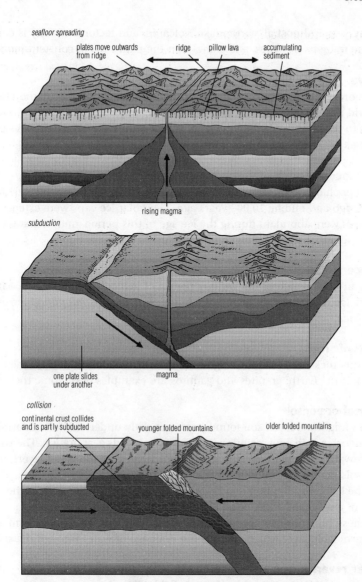

Constructive and destructive action in plate tectonics. (top) Seafloor spreading. The upwelling of magma forces apart the crust plates, producing new crust at the joint. Rapid extrusion of magma produces a domed ridge; more gentle spreading produces a central valley. (middle) The drawing downwards of an oceanic plate beneath a continent produces a range of volcanic fold mountains parallel to the plate edge. (bottom) Collision of continental plates produces immense fold mountains, such as the Himalayas.

Younger mountains are found near the coast with older ranges inland. The plates of the Earth's lithosphere are always changing in size and shape of each plate as material is added at constructive margins and removed at destructive margins. The process is extremely slow, but it means that the tectonic history of the Earth cannot be traced back further than about 200 million years.

Glossary

many years. Almost all earthquake, volcanic, and tectonic activity is confined to the margins of plates, and shows that the plates are in constant motion.

playa
temporary lake in a region of interior drainage. Such lakes are common features in arid desert basins fed by intermittent streams. The streams bring dissolved salts to the lakes, and when the lakes shrink during dry spells, the salts precipitate as evaporite deposits.

Pleistocene Epoch
first epoch of the Quaternary period of geological time, beginning 1.64 million years ago and ending 10,000 years ago. The polar ice caps were extensive and glaciers were abundant during the ice age of this period, and humans evolved into modern *Homo sapiens sapiens* about 100,000 years ago.

Pliocene Epoch
fifth and last epoch of the Tertiary period of geological time, 5.2–1.64 million years ago. The earliest hominid, the humanlike ape *Australopithecines*, evolved in Africa.

plutonic rock
igneous rock derived from magma that has cooled and solidified deep in the crust of the Earth; granites and gabbros are examples of plutonic rocks.

podzol or podsol
type of light-coloured soil found predominantly under coniferous forests and on moorlands in cool regions where rainfall exceeds evaporation. The constant downward movement of water leaches nutrients from the upper layers, making podzols poor agricultural soils.

The leaching of minerals such as iron, lime, and alumina leads to the formation of a bleached zone, often also depleted of clay.

These minerals can accumulate lower down the soil profile to form a hard, impermeable layer which restricts the drainage of water through the soil.

polar reversal or magnetic reversal
change in polarity of Earth's magnetic field. Like all magnets, Earth's magnetic field has two opposing regions, or poles, positioned approximately near geographical North and South Poles. During a period of normal polarity the region of attraction corresponds with the North Pole. Today, a compass needle, like other magnetic materials, aligns itself parallel to the magnetizing force and points to the North Pole. During a period of reversed polarity, the region of attraction would change to the South Pole and the needle of a compass would point south.

Studies of the magnetism retained in rocks at the time of their formation (like little compasses frozen in time) have shown that the polarity of the magnetic field has reversed repeatedly throughout geological time.

The reason for polar reversals is not known. Although the average time between reversals over the last 10 million years has been 250,000 years, the rate of reversal has changed continuously over geological time. The most recent reversal was 780,000 years ago; scientists have no way of predicting when the next reversal will occur. The reversal process probably takes a few thousand years. Dating rocks using distinctive sequences of magnetic reversals is called magnetic stratigraphy.

polder
area of flat reclaimed land that used to be covered by a river, lake, or the sea. Polders have been artificially drained and protected from flooding by building dykes. They are common in the Netherlands, where the total land area has been increased by nearly one-fifth since AD 1200. Such schemes as the Zuider Zee project have provided some of the best agricultural land in the country.

Precambrian
time from the formation of Earth (4.6 billion years ago) up to 570 million years ago. Its boundary with the succeeding Cambrian period marks the time when animals first developed hard outer parts (exoskeletons) and so left abundant fossil remains. It comprises about 85% of geological time and is divided into two eons: the Archaean and the Proterozoic.

precipitation
in meteorology, water that falls to the Earth from the atmosphere. It is part of the hydrological cycle. Forms of precipitation include rain, snow, sleet, hail, dew, and frost.

The amount of precipitation in any one area depends on climate weather, and phenomena like trade winds and ocean currents. The cyclical change in the Peru current off the coasts of Ecuador and Peru, El Niño, causes dramatic shifts in the amount of precipitation in South and Central America and throughout the Pacific region.

Precipitation can also be influenced by people. In urban areas dust, smoke, and other particulate pollution that comprise **condensation nuclei** cause water in the air to condense more readily. Fog is one example. Precipitation also can react chemically with air-borne pollutants producing acid rain.

Proterozoic Eon
eon of geological time, 3.5 billion to 570 million years ago, the second division of the Precambrian. It is defined as the time of simple life, since many rocks dating from this eon show traces of biological activity, and some contain the fossils of bacteria and algae.

pyroclastic deposit
deposit made up of fragments of rock, ranging in size from fine ash to large boulders, ejected during an explosive volcanic eruption.

Glossary

Quaternary Period
period of geological time from 1.64 million years ago through to the present. It is divided into the Pleistocene and Holocene epochs.

quartz
crystalline form of silica SiO_2, one of the most abundant minerals of the Earth's crust (12% by volume). Quartz occurs in many different kinds of rock, including sandstone and granite. It ranks 7 on the Mohs scale of hardness and is resistant to chemical or mechanical breakdown. Quartzes vary according to the size and purity of their crystals. Crystals of pure quartz are coarse, colourless, transparent, show no cleavage, and fracture unevenly; this form is usually called rock crystal. Impure coloured varieties, often used as gemstones, include agate, citrine quartz, and amethyst. Quartz is also used as a general name for the cryptocrystalline and noncrystalline varieties of silica, such as chalcedony, chert, and opal. Quartz is used in ornamental work and industry, where its reaction to electricity makes it valuable in electronic instruments. Quartz can also be made synthetically.

remote sensing
process of making observations of a planetary surface or atmosphere from far away, for example from an aeroplane or satellite. Remote sensing usually refers to gathering data using the electromagnetic spectrum (such as visible light, ultraviolet light, and infrared light).

Remote sensing is most commonly taken to refer to the process of photographing the Earth's surface with orbiting satellites. With a simple aerial, receiver, and software it is possible to download images straight on to a personal computer – helping amateur meteorologists, for example, to make weather forecasts.

resources
materials that can be used to satisfy human needs. Because human needs are diverse and extend from basic physical requirements, such as food and shelter, to ill-defined aesthetic needs, resources encompass a vast range of items. The intellectual resources of a society – its ideas and technologies – determine which aspects of the environment meet that society's needs, and therefore become resources. For example, in the 19th century, uranium was used only in the manufacture of coloured glass. Today, with the advent of nuclear technology, it is a military and energy resource. Resources are often categorized into ***human resources***, such as labour, supplies, and skills, and ***natural resources***, such as climate, fossil fuels, and water. Natural resources are divided into nonrenewable resources and renewable resources.

Nonrenewable resources include minerals such as coal, copper ores, and diamonds, which exist in strictly limited quantities. Once consumed they will not be replenished within the time span of human history. In contrast, water supplies, timber, food crops, and similar resources can, if managed properly, provide a steady yield virtually forever; they are therefore replenishable or renewable resources. Inappropriate use of renewable resources can lead to their destruction, as for example the cutting down of rainforests, with secondary effects, such as the decrease in oxygen and the increase in carbon dioxide and the ensuing greenhouse effect. Some renewable resources, such as wind or solar energy, are continuous; supply is largely independent of people's actions.

Demands for resources made by rich nations are causing concern that the present and future demands of industrial societies cannot be sustained for more than a century or two, and that this will be at the expense of the Third World and the global environment. Other authorities believe that new technologies will emerge, enabling resources that are now of little importance to replace those being exhausted.

ria
long narrow sea inlet, usually branching and surrounded by hills. A ria is deeper and wider towards its mouth, unlike a fjord. It is formed by the flooding of a river valley due to either a rise in sea level or a lowering of a landmass.

There are a number of rias in the UK – for example, in Salcombe and Dartmouth in Devon.

Richter scale
quantitative scale of earthquake magnitude based on measurement of seismic waves, used to indicate the magnitude of an earthquake at its epicentre. The magnitude of an earthquake differs from its intensity, measured by the Mercalli scale, which is qualitative and varies from place to place for the same earthquake. The scale is named after US seismologist Charles Richter.

An earthquake's magnitude is a function of the total amount of energy released, and each point on the Richter scale represents a thirtyfold increase in energy over the previous point. The greatest earthquake ever recorded, in 1920 in Gansu, China, measured 8.6 on the Richter scale.

rift valley
valley formed by the subsidence of a block of the Earth's crust between two or more parallel faults. Rift valleys are steep-sided and form where the crust is being pulled apart, as at ocean ridges, or in the Great Rift Valley of East Africa.

roche moutonnée
outcrop of tough bedrock having one smooth side and one jagged side, found on the floor of a glacial trough (U-shaped valley). It may be up to 40 m/130 ft high. A roche moutonnée is a feature of glacial erosion – as a glacier moved over its surface, ice and debris eroded its upstream side by corrasion, smoothing it and creating long scratches or striations. On the sheltered downstream side fragments of rock were plucked away by the ice, causing it to become steep and jagged.

Glossary

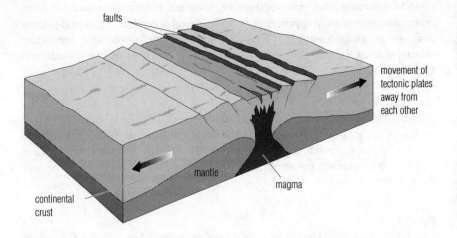

The subsidence of rock resulting from two or more parallel rocks moving apart is known as a graben. When this happens on a large scale, with tectonic plates moving apart, a rift valley is created.

rock
solid piece of the Earth or any other inorganic body in the Solar System. Rocks are composed of minerals or materials of organic origin. There are three basic types of rocks: igneous, sedimentary, or metamorphic rocks. Because rocks are composed of a combination (or aggregate) of minerals, the property of a rock will depend on its components. Where deposits of economically valuable minerals occur they are termed ores. As a result of weathering, rock breaks down into very small particles that combine with organic materials from plants and animals to form soil. In geology the term 'rock' can also include unconsolidated materials such as sand, mud, clay, and peat.

Igneous rock is formed by the cooling and solidification of magma, the molten rock material that originates in the lower part of the Earth's crust, or mantle, where it reaches temperatures as high as 1,000°C. The rock may form on or below the Earth's surface and is usually crystalline in texture. Larger crystals are more common in rocks such as granite which have cooled slowly within the Earth's crust; smaller crystals form in rocks such as basalt which have cooled more rapidly on the surface. Because of their acidic composition, igneous rocks such as granite are particularly susceptible to acid rain.

Sedimentary rocks are formed by the compression of particles deposited by water, wind, or ice. They may be created by the erosion of older rocks, the deposition of organic materials, or they may be formed from chemical precipitates. For example, sandstone is derived from sand particles, limestone from the remains of sea creatures, and gypsum is precipitated from evaporating sea water. Sedimentary rocks are typically deposited in distinct layers or strata and many contain fossils.

Metamorphic rocks are formed through the action of high pressure or heat on existing igneous or sedimentary rocks, causing changes to the composition, structure, and texture of the rocks. For example, marble is formed by the effects of heat and pressure on limestone, while granite may be metamorphosed into gneiss, a coarse-grained foliated rock.

ruby
red transparent gem variety of the mineral corundum Al_2O_3, aluminium oxide. Small amounts of chromium oxide, Cr_2O_3, substituting for aluminium oxide, give ruby its colour. Natural rubies are found mainly in Myanmar (Burma), but rubies can also be produced artificially and such synthetic stones are used in lasers.

salinization
accumulation of salt in water or soil; it is a factor in desertification.

salt, common or sodium chloride
NaCl white crystalline solid, found dissolved in sea water and as rock salt (the mineral halite) in large deposits and salt domes. Common salt is used extensively in the food industry as a preservative and for flavouring, and in the chemical industry in the making of chlorine and sodium.

salt marsh
wetland with halophytic vegetation (tolerant to sea water). Salt marshes develop around estuaries and on the sheltered side of sand and shingle spits. Salt marshes usually have a network of creeks and drainage channels by which tidal waters enter and leave the marsh.

Typical plants of European salt marshes include salicornia, or saltwort, which has fleshy leaves like a succulent; sea lavender, sea pink, and sea aster. Geese such as brent, greylag, and bean are frequent visitors to salt marshes in winter, feeding on plant material.

San Andreas fault
geological fault stretching for 1,125 km/700 mi northwest–southeast through the state of California, USA. It marks a conservative plate margin, where two plates slide past each other (see plate tectonics).

Friction is created as the coastal Pacific plate moves northwest, rubbing against the American continental plate, which is moving slowly southeast. The relative movement is only about 5 cm/2 in a year, which means that Los Angeles will reach San Francisco's latitude in 10 million years. The friction caused by the tectonic movement gives rise to frequent, destructive earthquakes. For example, in 1906 an earthquake originating from the fault almost destroyed San Francisco and killed about 700 people.

Glossary

sand

loose grains of rock, sized 0.0625–2.00 mm/0.0025–0.08 in in diameter, consisting most commonly of quartz, but owing their varying colour to mixtures of other minerals. Sand is used in cement-making, as an abrasive, in glass-making, and for other purposes.

sandstone

sedimentary rocks formed from the consolidation of sand, with sand-sized grains (0.0625–2 mm/0.0025–0.08 in) in a matrix or cement. Their principal component is quartz. Sandstones are commonly permeable and porous, and may form freshwater aquifers. They are mainly used as building materials.

Santa Ana

periodic warm Californian wind.

sapphire

deep-blue, transparent gem variety of the mineral corundum Al_2O_3, aluminium oxide. Small amounts of iron and titanium give it its colour. A corundum gem of any colour except red (which is a ruby) can be called a sapphire; for example, yellow sapphire.

satellite image

image of the Earth or any other planet obtained from instruments on a satellite. Satellite images can provide a variety of information, including vegetation patterns, sea surface temperature, weather, and geology.

Landsat 4, launched in 1982, orbits at 705 km/438 mi above the Earth's surface. It completes nearly 15 orbits per day, and can survey the entire globe in 16 days. The instruments on Landsat scan the planet's surface and record the brightness of reflected light. The data is transmitted back to Earth and translated into a satellite image.

scarp and dip

two slopes which comprise an escarpment. The scarp is the steep slope and the dip is the gentle slope. Such a feature is common when sedimentary rocks are uplifted, folded, or eroded, the scarp slope cuts across the bedding planes of the sedimentary rock whilst the dip slope follows the direction of the strata. An example is Salisbury Crags in Edinburgh, Scotland.

schist

metamorphic rock containing mica or another platy (flat and flakey) or elongate mineral, whose crystals are aligned to give a foliation (planar texture) known as schistosity. Schist may contain additional minerals such as garnet.

scree

pile of rubble and sediment that collects at the foot of a mountain range or cliff. The rock fragments that form scree are usually broken off by the action of frost (freeze–thaw weathering).

With time, the rock waste builds up into a heap or sheet of rubble that may eventually bury even the upper cliffs, and the growth of the scree then stops. Usually, however, erosional forces remove the rock waste so that the scree stays restricted to lower slopes.

seafloor spreading

growth of the ocean crust outwards (sideways) from ocean ridges. The concept of seafloor spreading has been combined with that of continental drift and incorporated into plate tectonics.

Seafloor spreading was proposed in 1960 by US geologist Harry Hess, based on his observations of ocean ridges and the relative youth of all ocean beds. In 1963, British geophysicists Fred Vine and Drummond Matthews observed that the floor of the Atlantic Ocean was made up of rocks that could be arranged in strips, each strip being magnetized either normally or reversely (due to changes in the Earth's polarity when the North Pole becomes the South Pole and vice versa, termed polar reversal). These strips were parallel and formed identical patterns on both sides of the ocean ridge. The implication was that each strip was formed at some stage in geological time when the magnetic field was polarized in a certain way. The seafloor magnetic-reversal patterns could be matched to dated magnetic reversals found in terrestrial rock. It could then be shown that new rock forms continuously and spreads away from the ocean ridges, with the oldest rock located farthest away from the midline. The observation was made independently in 1963 by Canadian geologist Lawrence Morley, studying an ocean ridge in the Pacific near Vancouver Island.

Confirmation came when sediments were discovered to be deeper further away from the oceanic ridge, because the rock there had been in existence longer and had had more time to accumulate sediment.

sea level

average height of the surface of the oceans and seas measured throughout the tidal cycle at hourly intervals and computed over a 19-year period. It is used as a datum plane from which elevations and depths are measured.

Factors affecting sea level include temperature of seawater (warm water is less dense and therefore takes up a greater volume than cool water) and the topography of the ocean floor.

secular variation

changes in the position of Earth's magnetic poles measured with respect to geographical positions, such as the North Pole, throughout geological time.

sedimentary rock

rock formed by the accumulation and cementation of deposits that have been laid down by water, wind, ice, or gravity. Sedimentary rocks cover more than two-thirds of the Earth's surface and comprise three major categories: clastic, chemically precipitated, and organic (or biogenic). Clastic sediments are the

Glossary

largest group and are composed of fragments of pre-existing rocks; they include clays, sands, and gravels.

Chemical precipitates include some limestones and evaporated deposits such as gypsum and halite (rock salt). Coal, oil shale, and limestone made of fossil material are examples of organic sedimentary rocks.

Most sedimentary rocks show distinct layering (stratification), because they are originally depositied as essentially horizontal layers.

seiche
pendulous movement seen in large areas of water resembling a tide. It was originally observed on Lake Geneva and is created either by the wind, earth tremors, or other atmospheric phenomena.

seismic gap theory
theory that along faults that are known to be seismically active, or in regions of high seismic activity, the locations that are more likely to experience an earthquake in the relatively near future are those that have not shown seismic activity for some time. When records of past earthquakes are studied and plotted onto a map, it becomes possible to identify *seismic gaps* along a fault or plate margin. According to the theory, an area that has not had an earthquake for some time will have a great deal of stress building up, which must eventually be released, causing an earthquake.

Although the seismic gap theory can suggest areas that are likely to experience an earthquake, it does not enable scientists to predict when that earthquake will occur.

seismic wave
energy wave generated by an earthquake or an artificial explosion. There are two types of seismic waves: *body waves* that travel through the Earth's interior and *surface waves* that travel through the surface layers of the crust and can be felt as the shaking of the ground, as in an earthquake.

Body waves
There are two types of body waves: P-waves and S-waves, so-named because they are the primary and secondary waves detected by a seismograph. *P-waves* are longitudinal waves (wave motion in the direction the wave is travelling), whose compressions and rarefactions resemble those of a sound wave. *S-waves* are transverse waves or shear waves, involving a back-and-forth shearing motion at right angles to the direction the wave is travelling.

Because liquids have no resistance to shear and cannot sustain a shear wave, S-waves cannot travel through liquid material. The Earth's outer core is believed to be liquid because S-waves disappear at the mantle–core boundary, while P-waves do not.

Surface waves
Surface waves travel in the surface and subsurface layers of the crust.

Rayleigh waves travel along the free surface (the uppermost layer) of a solid material. The motion of particles is elliptical, like a water wave, creating the rolling motion often felt during an earthquake. *Love waves* are transverse waves trapped in a subsurface layer due to different densities in the rock layers above and below. They have a horizontal side-to-side shaking motion transverse (at right angles) to the direction the wave is travelling.

seismogram or seismic record
trace, or graph, of ground motion over time, recorded by a seismograph. It is used to determine the magnitude and duration of an earthquake.

seismograph
instrument used to record ground motion. A heavy inert weight is suspended by a spring and attached to this is a pen that is in contact with paper on a rotating drum. During an earthquake the instrument frame and drum move, causing the pen to record a zigzag line on the paper; held steady by inertia, the pen does not move.

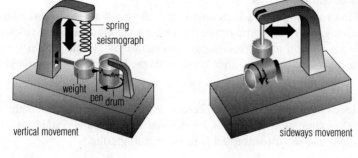

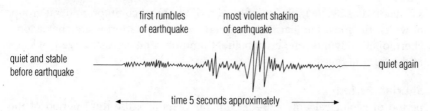

A seismogram, or recording made by a seismograph. Such recordings are used to study earthquakes and in prospecting.

seismology
study of earthquakes, the seismic waves they produce, the processes that cause them, and the effects they have. By examining the global pattern of waves produced by an earthquake, seismologists can deduce the nature of the materials through which they have passed. This leads to an understanding of the Earth's internal structure.

Glossary

On a smaller scale, artificial earthquake waves, generated by explosions or mechanical vibrators, can be used to search for subsurface features in, for example, oil or mineral exploration. Earthquake waves from underground nuclear explosions can be distinguished from natural waves by their shorter wavelength and higher frequency.

selva
equatorial rainforest, such as that in the Amazon basin in South America.

shale
fine-grained and finely layered sedimentary rock composed of silt and clay. It is a weak rock, splitting easily along bedding planes to form thin, even slabs (by contrast, mudstone splits into irregular flakes). Oil shale contains kerogen, a solid bituminous material that yields petroleum when heated.

shield
alternative name for craton, the ancient core of a continent.

shield volcano
broad flat volcano formed at a constructive margin between tectonic plates or over a hot spot. The magma (molten rock) associated with shield volcanoes is usually basalt-thin and free-flowing. An example is Mauna Loa in Hawaii. A composite volcano, on the other hand, is formed at a destructive margin.

sial
substance of the Earth's continental crust, as distinct from the sima of the ocean crust. The name, now used rarely, is derived from *si*lica and *al*umina, its two main chemical constituents. Sial is often rich in granite.

silica
sj21ilicon dioxide, SiO_2, the composition of the most common mineral group, of which the most familiar form is quartz. Other silica forms are chalcedony, chert, opal, tridymite, and cristobalite. Common sand consists largely of silica in the form of quartz.

Silurian Period
period of geological time 439–409 million years ago, the third period of the Palaeozoic era. Silurian sediments are mostly marine and consist of shales and limestone. Luxuriant reefs were built by coral-like organisms. The first land plants began to evolve during this period, and there were many ostracoderms (armoured jawless fishes). The first jawed fishes (called acanthodians) also appeared.

sima
substance of the Earth's oceanic crust, as distinct from the sial of the continental crust. The name, now used rarely, is derived from *si*lica and *ma*gnesia, its two main chemical constituents.

sirocco
hot, normally dry and dust-laden wind that blows from the deserts of North Africa across the Mediterranean into southern Europe. It occurs mainly in the spring. The name 'sirocco' is also applied to any hot oppressive wind.

slate
fine-grained, usually grey metamorphic rock that splits readily into thin slabs along its cleavage planes. It is the metamorphic equivalent of shale.

snout
front end of a glacier, representing the furthest advance of the ice at any one time. Deep cracks, or crevasses, and ice falls are common.

Because the snout is the lowest point of a glacier it tends to be affected by the warmest weather. Considerable melting takes place, and so it is here that much of the rocky material carried by the glacier becomes deposited. Material dumped just ahead of the snout may form a terminal moraine.

The advance or retreat of the snout depends upon the glacier budget – the balance between accumulation (the addition of snow and ice to the glacier) and ablation (their loss by melting and evaporation).

soil
loose covering of broken rocky material and decaying organic matter overlying the bedrock of the Earth's surface. It is comprised of minerals, organic matter (called humus) derived from decomposed plants and organisms, living organisms, air, and water. Soils differ according to climate, parent material, rainfall, relief of the bedrock, and the proportion of organic material. The study of soils is *pedology*.

A soil can be described in terms of its *soil profile*, that is, a vertical cross-section from ground-level to the bedrock on which the soils sits. The profile is divided into layers called horizons. The A horizon, or topsoil, is the uppermost layer, consisting primarily of humus and living organisms and some mineral material. Most soluble material has been leached from this layer or washed down to the B horizon. The B horizon, or subsoil, is the layer where most of the nutrients accumulate and is enriched in clay minerals. The C horizon is the layer of weathered parent material at the base of the soil.

Two common soils are the podzol and the *chernozem* soil. The podzol is common in coniferous forest regions where precipitation exceeds evaporation. The A horizon consists of a very thin litter of organic material producing a poor humus. Needles take a long time to decompose. The relatively heavy precipitation causes leaching of minerals, as nutrients are washed downwards.

Chernozem soils are found in grassland regions, where evaporation exceeds precipitation. The A horizon is rich in humus due to decomposition of a thick litter of dead grass at the surface. Minerals and moisture migrate upward due to evaporation, leaving the B and A horizons enriched.

The organic content of soil is widely variable, ranging from zero in some desert soils to almost 100% in peats.

Glossary

Soils influence the type of agriculture employed in a particular region – light well-drained soils favour arable farming, whereas heavy clay soils give rise to lush pasture land.

soil creep
gradual movement of soil down a slope in response to gravity. This eventually results in a mass downward movement of soil on the slope.

Evidence of soil creep includes the formation of terracettes (steplike ridges along the hillside), leaning walls and telegraph poles, and trees that grow in a curve to counteract progressive leaning.

soil erosion
wearing away and redistribution of the Earth's soil layer. It is caused by the action of water, wind, and ice, and also by improper methods of agriculture. If unchecked, soil erosion results in the formation of deserts (desertification). It has been estimated that 20% of the world's cultivated topsoil was lost between 1950 and 1990.

If the rate of erosion exceeds the rate of soil formation (from rock and decomposing organic matter), then the land will become infertile. The removal of forests (deforestation) or other vegetation often leads to serious soil erosion, because plant roots bind soil, and without them the soil is free to wash or blow away, as in the US dust bowl. The effect is worse on hillsides, and there has been devastating loss of soil where forests have been cleared from mountainsides, as in Madagascar.

Improved agricultural practices such as contour ploughing are needed to combat soil erosion. Windbreaks, such as hedges or strips planted with coarse grass, are valuable, and organic farming can reduce soil erosion by as much as 75%.

Soil degradation and erosion are becoming as serious as the loss of the rainforest. It is estimated that more than 10% of the world's soil lost a large amount of its natural fertility during the latter half of the 20th century. Some of the worst losses are in Europe, where 17% of the soil is damaged by human activity such as mechanized farming and fallout from acid rain. Mexico and Central America have 24% of soil highly degraded, mostly as a result of deforestation.

solifluction
downhill movement of topsoil that has become saturated with water. Solifluction is common in periglacial environments (those bordering glacial areas) during the summer months, when the frozen topsoil melts to form an unstable soggy mass. This may then flow slowly downhill under gravity to form a *solifluction lobe* (a tonguelike feature).

Solifluction material, or head, is found at the bottom of chalk valleys in southern England; it is partly responsible for the rolling landscape typical of chalk scenery.

solution or dissolution

process by which the minerals in a rock are dissolved in water. Solution is one of the processes of erosion as well as weathering (in which the dissolution of rock occurs without transport of the dissolved material). An example of this is when weakly acidic rainfall dissolves calcite.

speleology

scientific study of caves, their origin, development, physical structure, flora, fauna through exploration, mapping, photography, folklore, cave-diving, and rescue work. *Potholing*, which involves following the course of underground rivers or streams, has become a popular sport.

Speleology first developed in France in the late 19th century, where the Société de Spéléologie was founded in 1895.

spit

ridge of sand or shingle projecting from the land into a body of water. It is formed by the interruption of longshore drift due to wave interaction with tides, currents, or a bend in the coastline. The consequent decrease in wave energy causes more material to be deposited than is transported down the coast, building up a finger of sand that points in the direction of the longshore drift. Deposition in the brackish water behind a spit may result in the formation of a salt marsh.

sprite

rare thunderstorm-related luminous flash. Sprites occur in the mesosphere, at altitudes of 50–90 km/30–55 mi. They are electrical, like lightning, and arise when the electrical field that occurs between the thunder cloud top and the ionosphere (ionized layer of the Earth's atmosphere) draws electrons upwards

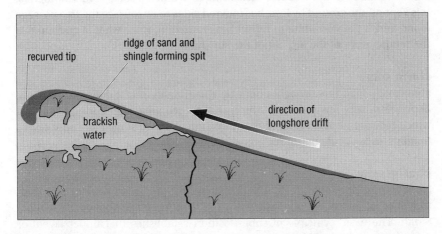

Longshore drift carries sand and shingle up coastlines. Deposited material gradually builds up over time at headlands forming a new stretch of land called a spit. A spit that extends across a bay is known as a bar.

from the cloud. If the air is thin and this field is strong the electrons accelarate rapidly, transferring kinetic energy to molecules they collide with. The excited molecules then discharge this energy as a light flash.

stack
isolated pillar of rock that has become separated from a headland by coastal erosion. It is usually formed by the collapse of an arch. Further erosion will reduce it to a stump, which is exposed only at low tide.

Examples of stacks in the UK are the Needles, off the Isle of Wight, which are formed of chalk.

stalactite and stalagmite
cave structures formed by the deposition of calcite dissolved in groundwater. *Stalactites* grow downwards from the roofs or walls and can be icicle-shaped, straw-shaped, curtain-shaped, or formed as terraces.

Stalagmites grow upwards from the cave floor and can be conical, fir-cone-shaped, or resemble a stack of saucers. Growing stalactites and stalagmites may meet to form a continuous column from floor to ceiling.

Stalactites are formed when groundwater, hanging as a drip, loses a proportion of its carbon dioxide into the air of the cave. This reduces the amount of calcite that can be held in solution, and a small trace of calcite is deposited. Successive drips build up the stalactite over many years. In stalagmite formation the calcite comes out of the solution because of agitation – the shock of a drop of water hitting the floor is sufficient to remove some calcite from the drop. The different shapes result from the splashing of the falling water.

Stevenson screen
box designed to house weather-measuring instruments such as thermometers. It is kept off the ground by legs, has louvred sides to encourage the free passage of air, and is painted white to reflect heat radiation, since what is measured is the temperature of the air, not of the sunshine.

storm surge
abnormally high tide brought about by a combination of a deep atmospheric depression (very low pressure) over a shallow sea area, high spring tides, and winds blowing from the appropriate direction. A storm surge can cause severe flooding of lowland coastal regions and river estuaries.

stratigraphy
branch of geology that deals with sedimentary rock layers (strata) and their sequence of formation. Its basis was developed by English geologist William Smith. The basic principle of superposition establishes that upper layers or deposits accumulated later in time than the lower ones.

Stratigraphy involves both the investigation of sedimentary structures to determine past environments represented by rocks, and the study of fossils for identifying and dating particular beds of rock. A body of rock strata with a set of unifying characteristics indicative of an environment is called a facies.

Stratigraphic units can be grouped in terms of time or lithology (rock type). Strata that were deposited at the same time belong to a single *chronostratigraphic unit* but need not be the same lithology. Strata of a specific lithology can be grouped into a *lithostratigraphic unit* but are not necessarily the same age.

Stratigraphy in the interpretation of archaeological excavations provides a relative chronology for the levels and the artefacts within rock beds. It is the principal means by which the context of archaeological deposits is evaluated.

stratosphere
that part of the atmosphere 10–40 km/6–25 mi from the Earth's surface, where the temperature slowly rises from a low of −55°C/−67°F to around 0°C/32°F. The air is rarefied and at around 25 km/15 mi much ozone is concentrated.

strip mining or open-pit mining
mining from the surface rather than by tunneling underground. Coal, iron ore, and phosphates are often extracted by strip mining. Often the mineral deposit is covered by soil, which must first be stripped off, usually by large machines such as walking draglines and bucket-wheel excavators. The ore deposit is then broken up by explosives and collected from the surface.

stromatolite
mound produced in shallow water by mats of algae that trap mud particles. Another mat grows on the trapped mud layer and this traps another layer of mud and so on. The stromatolite grows to heights of a metre or so. They are uncommon today but their fossils are among the earliest evidence for living things – over 2,000 million years old.

stump
low outcrop of rock formed by the erosion of a coastal stack. Unlike a stack, which is exposed at all times, a stump is exposed only at low tide. Eventually it will be worn away completely, leaving a wave-cut platform.

subduction zone
region where two plates of the Earth's rigid lithosphere collide, and one plate descends below the other into the weaker asthenosphere. Subduction results in the formation of ocean trenches, most of which encircle the Pacific Ocean.

Ocean trenches are usually associated with volcanic island arcs and deep-focus earthquakes (more than 300 km/185 mi below the surface), both the result of disturbances caused by the plate subduction.

swallow hole or swallet
hole, often found in limestone areas, through which a surface stream disappears underground. It will usually lead to an underground network of caves. Gaping Gill in North Yorkshire is an example.

syncline
fold in the rocks of the Earth's crust in which the layers or beds dip inwards, thus forming a trough-like structure with a sag in the middle. The opposite structure, with the beds arching upwards, is an anticline.

synoptic chart
weather chart in which symbols are used to represent the weather conditions experienced over an area at a particular time. Synoptic charts appear on television and newspaper forecasts, although the symbols used may differ.

talc
$Mg_3Si_4O_{10}(OH)_2$, mineral, hydrous magnesium silicate. It occurs in tabular crystals, but the massive impure form, known as **steatite** or **soapstone**, is more common. It is formed by the alteration of magnesium compounds and is usually found in metamorphic rocks. Talc is very soft, ranked 1 on the Mohs scale of hardness. It is used in powdered form in cosmetics, lubricants, and as an additive in paper manufacture.

tectonics
study of the movements of rocks on the Earth's surface. On a small scale tectonics involves the formation of folds and faults, but on a large scale plate tectonics deals with the movement of the Earth's surface as a whole.

terminal moraine
linear, slightly curved ridge of rocky debris deposited at the front end, or snout, of a glacier. It represents the furthest point of advance of a glacier, being formed when deposited material (till), which was pushed ahead of the snout as it advanced, became left behind as the glacier retreated.

A terminal moraine may be hundreds of metres in height; for example, the Franz Joseph glacier in New Zealand has a terminal moraine that is over 400 m/1,320 ft high.

terrane
tract of land with a distinct geological character. The term *exotic terrane* is commonly used to describe a rock mass that has a very different history from others near by. The exotic terranes of the Western Cordillera of North America represent old island chains that have been brought to the North American continent by the movements of plate tectonics, and welded to its edge.

terrigenous
derived from or pertaining to the land. River sediment composed of weathered rock material and deposited near the mouth of the river on the ocean's continental shelf (the shallow ledge extending out from the continent) is called *terrigenous sediment*.

Tertiary Period
period of geological time 65–1.64 million years ago, divided into five epochs: Palaeocene, Eocene, Oligocene, Miocene, and Pliocene. During the Tertiary period, mammals took over all the ecological niches left vacant by the extinction of the dinosaurs, and became the prevalent land animals. The continents took on their present positions, and climatic and vegetation zones as we know them became established. Within the geological time column the Tertiary follows the Cretaceous period and is succeeded by the Quaternary period.

Tethys Sea
sea that in the Mesozoic era separated Laurasia from Gondwanaland. The formation of the Alpine fold mountains caused the sea to separate into the Mediterranean, the Black, the Caspian, and the Aral seas.

thermosphere
layer in the Earth's atmosphere above the mesosphere and below the exosphere. Its lower level is about 80 km/50 mi above the ground, but its upper level is undefined. The ionosphere is located in the thermosphere. In the thermosphere the temperature rises with increasing height to several thousand degrees Celsius. However, because of the thinness of the air, very little heat is actually present.

tidal wave
common name for a tsunami.

tide
rhythmic rise and fall of the sea level in the Earth's oceans and their inlets and estuaries due to the gravitational attraction of the Moon and, to a lesser extent, the Sun, affecting regions of the Earth unequally as it rotates. Water on the side of the Earth nearest the Moon feels the Moon's pull and accumulates directly below it producing high tide.

High tide occurs at intervals of 12 hr 24 min 30 sec. The maximum high tides, or spring tides, occur at or near new and full Moon when the Moon and Sun are in line and exert the greatest combined gravitational pull. Lower high tides, or neap tides, occur when the Moon is in its first or third quarter and the Moon and Sun are at right angles to each other.

Gravitational tides – the pull of nearby groups of stars – have been observed to affect the galaxies.

topaz
mineral, aluminium fluorosilicate, $Al_2(F_2SiO_4)$. It is usually yellow, but pink if it has been heated, and is used as a gemstone when transparent. It ranks 8 on the Mohs scale of hardness.

Glossary

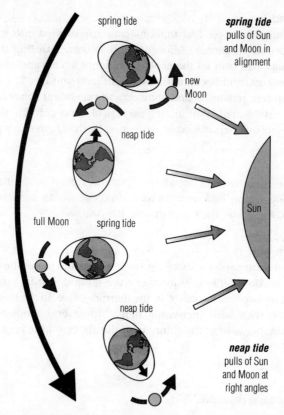

The gravitational pull of the Moon is the main cause of the tides. Water on the side of the Earth nearest the Moon feels the Moon's pull and accumulates directly under the Moon. When the Sun and the Moon are in line, at new and full Moon, the gravitational pull of Sun and Moon are in line and produce a high spring tide. When the Sun and Moon are at right angles, lower neap tides occur.

tornado

extremely violent revolving storm with swirling, funnel-shaped clouds, caused by a rising column of warm air propelled by strong wind. A tornado can rise to a great height, but with a diameter of only a few hundred metres or less. Tornadoes move with wind speeds of 160–480 kph/100–300 mph, destroying everything in their path. They are common in the central USA and Australia.

A series of tornadoes killed 47 people, destroyed 2,000 homes, and caused $500 million worth of damage in Oklahoma, Nebraska, Kansas, and Texas in May 1999.

Triassic Period

period of geological time 245–208 million years ago, the first period of the Mesozoic era. The present continents were fused together in the form of the world continent Pangaea. Triassic sediments contain remains of early dinosaurs and other animals now extinct. By late Triassic times, the first mammals had evolved.

There was a mass extinction of 95% of plants at the end of the Triassic possibly caused by rising temperatures.

The climate was generally dry; desert sandstones are typical Triassic rocks.

troilite
FeS, probable mineral of the Earth's core, abundant in meteorites.

tropical cyclone
another term for hurricane.

troposphere
lower part of the Earth's atmosphere extending about 10.5 km/6.5 mi from the Earth's surface, in which temperature decreases with height to about −60°C/ −76°F except in local layers of temperature inversion. The *tropopause* is the upper boundary of the troposphere, above which the temperature increases slowly with height within the atmosphere. All of the Earth's weather takes place within the troposphere.

truncated spur
blunt-ended ridge of rock jutting from the side of a glacial trough, or valley. As a glacier moves down a river valley it is unable to flow around the interlocking spurs that project from either side, and so it erodes straight through them, shearing away their tips and forming truncated spurs.

tsunami
ocean wave generated by vertical movements of the sea floor resulting from earthquakes or volcanic activity or large submarine landslides. Unlike waves generated by surface winds, the entire depth of water is involved in the wave motion of a tsunami. In the open ocean the tsunami takes the form of several successive waves, rarely in excess of 1 m/3 ft in height but travelling at speeds of 650–800 kph/400–500 mph. In the coastal shallows tsunamis slow down and build up producing huge swells over 15 m/45 ft high in some cases and over 30 m/90 ft in rare instances. The waves sweep inland causing great loss of life and property.

Before each wave there may be a sudden withdrawal of water from the beach. Used synonymously with tsunami, the popular term 'tidal wave' is misleading: tsunamis are not caused by the gravitational forces that affect tides.

tufa or travertine
soft, porous, limestone rock, white in colour, deposited from solution from carbonate-saturated groundwater around hot springs and in caves.

tundra

region of high latitude almost devoid of trees, resulting from the presence of permafrost. The vegetation consists mostly of grasses, sedges, heather, mosses, and lichens. Tundra stretches in a continuous belt across northern North America and Eurasia. Tundra is also used to describe similar conditions at high altitudes.

The term was originally applied to the topography of part of northern Russia, but is now used for all such regions.

turquoise

mineral, hydrous basic copper aluminium phosphate, $CuAl_6(PO_4)_4(OH)_8 5H_2O$. Blue-green, blue, or green, it is a gemstone. Turquoise is found in Australia, Egypt, Ethiopia, France, Germany, Iran, Turkestan, Mexico, and southwestern USA. It was originally introduced into Europe through Turkey, from which its name is derived.

typhoon

violent revolving storm, a hurricane in the western Pacific Ocean.

unconformity

surface of erosion or nondeposition eventually overlain by younger sedimentary rock strata and preserved in the geologic record. A surface where the beds above and below lie at different angles is called an ***angular unconformity***. The boundary between older igneous or metamorphic rocks that are truncated by erosion and later covered by younger sedimentary rocks is called a ***nonconformity***.

uniformitarianism

principle that processes that can be seen to occur on the Earth's surface today are the same as those that have occurred throughout geological time. For example, desert sandstones containing sand-dune structures must have been formed under conditions similar to those present in deserts today. The principle was formulated by Scottish geologists James Hutton and expounded by Charles Lyell.

U-shaped valley

another term for a glacial trough, a valley formed by a glacier.

varve

pair of thin sedimentary beds, one coarse and one fine, representing a cycle of thaw followed by an interval of freezing, in lakes of glacial regions.

Each couplet thus constitutes the sedimentary record of a year, and by counting varves in glacial lakes a record of absolute time elapsed can be determined. Summer and winter layers often are distinguished also by colour, with lighter layers representing summer deposition, and darker layers being the result of dark clay settling from water while the lake is frozen.

volcanic rock
another name for extrusive rock, igneous rock formed on the Earth's surface.

volcano
crack in the Earth's crust through which hot magma (molten rock) and gases well up. The magma is termed lava when it reaches the surface. A volcanic mountain, usually cone shaped with a crater on top, is formed around the opening, or vent, by the build-up of solidified lava and ashes (rock fragments). Most volcanoes arise on plate margins (see plate tectonics), where the movements of plates generate magma or allow it to rise from the mantle beneath. However, a number are found far from plate-margin activity, on 'hot spots' where the Earth's crust is thin. There are two main types of volcano: composite and shield.

Composite volcanoes, such as Stromboli and Vesuvius in Italy, are found at destructive plate margins (areas where plates are being pushed together), usually in association with island arcs and coastal mountain chains. The magma is mostly derived from plate material and is rich in silica. This makes a very stiff lava such as andesite, which solidifies rapidly to form a high, steep-sided volcanic mountain. The magma often clogs the volcanic vent, causing violent eruptions as the blockage is blasted free, as in the eruption of Mount St Helens, USA, in 1980. The crater may collapse to form a caldera.

Shield volcanoes, such as Mauna Loa in Hawaii, are found along the rift valleys and ocean ridges of constructive plate margins (areas where plates are

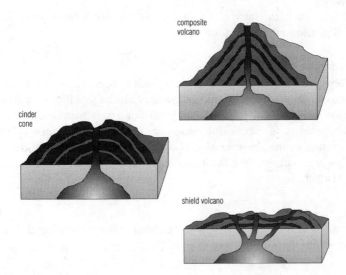

There are two main types of volcano, but three distinctive cone shapes. Composite volcanoes emit a stiff, rapidly solidifying lava which forms high, steep-sided cones. Volcanoes that regularly throw out ash build up flatter domes known as cinder cones. The lava from a shield volcano is not ejected violently, flowing over the crater rim forming a broad low profile.

moving apart), and also over hot spots. The magma is derived from the Earth's mantle and is quite free-flowing. The lava formed from this magma – usually basalt – flows for some distance over the surface before it sets and so forms broad low volcanoes. The lava of a shield volcano is not ejected violently but simply flows over the crater rim.

The type of volcanic activity is also governed by the age of the volcano. The first stages of an eruption are usually vigorous as the magma forces its way to the surface. As the pressure drops and the vents become established, the main phase of activity begins, composite volcanoes giving pyroclastic debris and shield volcanoes giving lava flows. When the pressure from below ceases, due to exhaustion of the magma chamber, activity wanes and is confined to the emission of gases and in time this also ceases. The volcano then enters a period of quiescence, after which activity may resume after a period of days, years, or even thousands of years. Only when the root zones of a volcano have been exposed by erosion can a volcano be said to be truly extinct.

Many volcanoes are submarine and occur along mid-ocean ridges. The chief terrestrial volcanic regions are around the Pacific rim (Cape Horn to Alaska); the central Andes of Chile (with the world's highest active volcano, Guallatiri, 6,063 m/19,892 ft); North Island, New Zealand; Hawaii; Japan; and Antarctica. There are more than 1,300 potentially active volcanoes on Earth. Volcanism has helped shape other members of the Solar System, including the Moon, Mars, Venus, and Jupiter's moon Io.

volcanology
study of volcanoes, the lava, rocks, and gases that erupt from them, and the geological phenomena that cause them.

V-shaped valley
river valley with a V-shaped cross-section. Such valleys are usually found near the source of a river, where the steeper gradient means that there is a great deal of corrasion (grinding away by rock particles) along the stream bed and erosion cuts downwards more than it does sideways. However, a V-shaped valley may also be formed in the lower course of a river when its powers of downward erosion become renewed by a fall in sea level, a rise in land level, or the capture of another river.

wadi
in arid regions of the Middle East, a steep-sided valley containing an intermittent stream that flows in the wet season.

water cycle or hydrological cycle
natural circulation of water through the upperpart of the Earth. It is a complex system involving a number of physical and chemical processes (such as evaporation, precipitation, and infiltration) and stores (such as rivers, oceans, and soil).

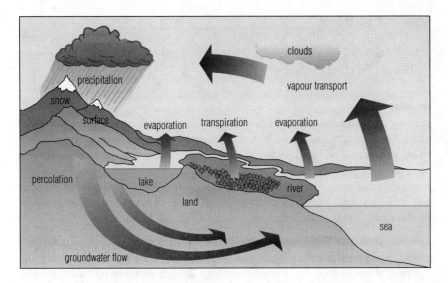

About one-third of the solar energy reaching the Earth is used in evaporating water. About 380,000 cubic km/95,000 cubic mi is evaporated each year. The entire contents of the oceans would take about one million years to pass through the water cycle.

Water is lost from the Earth's surface to the atmosphere by evaporation caused by the Sun's heat on the surface of lakes, rivers, and oceans, and through the transpiration of plants. This atmospheric water is carried by the air moving across the Earth, and **condenses** as the air cools to form clouds, which in turn deposit moisture on the land and sea as precipitation. The water that collects on land flows to the ocean overland – as streams, rivers, and glaciers – or through the soil (infiltration) and rock (groundwater). The boundary that marks the upper limit of groundwater is called the water table.

The oceans, which cover around 70% of the Earth's surface, are the source of most of the moisture in the atmosphere.

water table

upper level of groundwater (water collected underground in porous rocks). Water that is above the water table will drain downwards; a spring forms where the water table cuts the surface of the ground. The water table rises and falls in response to rainfall and the rate at which water is extracted, for example, for irrigation and industry.

In many irrigated areas the water table is falling due to the extraction of water. Below northern China, for example, the water table is sinking at a rate of 1 m/3 ft a year. Regions with high water tables and dense industrialization have problems with pollution of the water table. In the USA, New Jersey, Florida, and Louisiana have water tables contaminated by both industrial wastes and saline seepage from the ocean.

wave-cut platform
gently sloping rock surface found at the foot of a coastal cliff. Covered by water at high tide but exposed at low tide, it represents the last remnant of an eroded headland (see coastal erosion).

weather
variation of atmospheric conditions at any one place over a short period of time. Such conditions include humidity, precipitation, temperature, cloud cover, visibility, and wind. Weather differs from climate in that the latter is a composite of the average weather conditions of a locality or region over a long period of time (at least 30 years). Meteorology is the study of short-term weather patterns and data within a circumscribed area; climatology is the study of weather over longer timescales on a zonal or global basis.

weather forecasts Forecasts are based on current meteorological data, and predict likely weather for a particular area; they may be short-range (covering a period of one or two days), medium-range (five to seven days), or long-range (a month or so). Weather observations are made on an hourly basis at meteorological recording stations – there are more than 3,500 of these around the world. More than 140 nations participate in the exchange of weather data through the World Weather Watch programme, which is sponsored by the World Meteorological Organization (WMO), and information is distributed among the member nations by means of a worldwide communications network. Incoming data is collated at weather centres in individual countries and plotted on weather maps, or charts. The weather map uses internationally standardized symbols to indicate barometric pressure, cloud cover, wind speed and direction, precipitation, and other details reported by each recording station at a specific time. Points of equal atmospheric pressure are joined by lines called isobars and from these the position and movement of weather fronts and centres of high and low pressure can be extrapolated. The charts are normally compiled on a three-hourly or six-hourly basis – the main synoptic hours are midnight, 0600, 1200 and 1800 – and predictions for future weather are drawn up on the basis of comparisons between current charts and previous charts. Additional data received from weather balloons and satellites help to complete and corroborate the picture obtained from the weather map.

weathering
process by which exposed rocks are broken down on the spot by the action of rain, frost, wind, and other elements of the weather. It differs from erosion in that no movement or transportation of the broken-down material takes place. Two types of weathering are recognized: physical (or mechanical) and chemical. They usually occur together.

Physical weathering
This is the mechanical breakdown of rocks but involving no chemical change. Examples include such effects as freeze–thaw (the splitting of rocks by the

Glossary

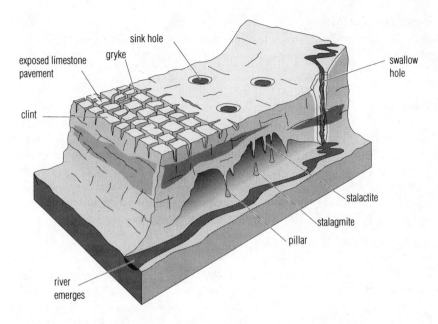

The physical weathering and erosion of a limestone landscape. The freezing and thawing of rain and its mild acidic properties cause cracks and joints to enlarge, forming limestone pavements, potholes, caves, and caverns.

alternate freezing and thawing of water trapped in cracks) and exfoliation, or onion-skin weathering (flaking caused by the alternate expansion and contraction of rocks in response to extreme changes in temperature).

Chemical weathering

Involving a chemical change in the rocks affected, the most common form is caused by rainwater that has absorbed carbon dioxide from the atmosphere and formed a weak carbonic acid. This then reacts with certain minerals in the rocks and breaks them down. Examples are the solution of caverns in limestone terrains, and the breakdown of feldspars in granite to form china clay or kaolin.

Although physical and chemical weathering normally occur together, in some instances it is difficult to determine which type is involved. For example, exfoliation, which produces rounded inselbergs in arid regions, such as Ayers Rock in central Australia, may be caused by the daily physical expansion and contraction of the surface layers of the rock in the heat of the Sun, or by the chemical reaction of the minerals just beneath the surface during the infrequent rains of these areas.

Appendix

Atmosphere: Composition

Gas	Symbol	Volume (%)	Role
nitrogen	N_2	78.08	cycled through human activities and through the action of micro-organisms on animal and plant waste
oxygen	O_2	20.94	cycled mainly through the respiration of animals and plants and through the action of photosynthesis
carbon dioxide	CO_2	0.03	cycled through respiration and photosynthesis in exchange reactions with oxygen. It is also a product of burning fossil fuels
argon	Ar	0.93	chemically inert and with only a few industrial uses
neon	Ne	0.0018	as argon
helium	He	0.0005	as argon
krypton	Kr	trace	as argon
xenon	Xe	trace	as argon
ozone	O_3	0.00006	a product of oxygen molecules split into single atoms by the Sun's radiation and unaltered oxygen molecules
hydrogen	H_2	0.00005	unimportant

Appendix

Mercalli Scale

The Mercalli scale is a measure of the intensity of an earthquake. It differs from the Richter scale, which measures magnitude. It is named after the Italian seismologist Giuseppe Mercalli (1850–1914). The scale shown here is the Modified Mercalli Intensity Scale, developed in 1931 by US seismologists Harry Wood and Frank Neumann.

Intensity value	Description
I	not felt except by a very few under especially favourable conditions
II	felt only by a few persons at rest, especially on upper floors of buildings
III	felt quite noticeably by persons indoors, especially on upper floors of buildings; many people do not recognize it as an earthquake; standing motor cars may rock slightly
IV	felt indoors by many, outdoors by a few persons during the day; at night, some awakened; dishes, windows, doors disturbed; walls make cracking sound; standing motor cars rock noticeably
V	felt by nearly everyone; many awakened; some dishes, windows broken; unstable objects overturned; pendulum clocks may stop
VI	felt by all; some heavy furniture moved; a few instances of fallen plaster; damage slight
VII	damage negligible in buildings of good design and construction; slight to moderate in well-built ordinary structures; considerable damage in poorly built or badly designed structures; some chimneys broken
VIII	damage slight in specially designed structures; considerable damage in ordinary substantial buildings with partial collapse; damage great in poorly built structures; fall of chimneys, factory stacks, columns, monuments, walls; heavy furniture overturned
IX	damage considerable in specially designed structures; damage great in substantial buildings, with partial collapse; buildings shifted off foundations
X	some well built wooden structures destroyed; most masonry and frame structures with foundations destroyed; rails bent
XI	few, if any (masonry) structures remain standing; bridges destroyed; rails bent greatly
XII	damage total; lines of sight and level are distorted; objects thrown into the air

The Richter Scale

The Richter scale is based on measurement of seismic waves, used to determine the magnitude of an earthquake at its epicentre. The Richter scale was named after US seismologist Charles Richter (1900–1985). The relative amount of energy released indicates the ratio of energy between earthquakes of different magnitude.

Magnitude	Relative amount of energy released	Examples	Year
1	1		
2	31		
3	960		
4	30,000	Carlisle, England (4.7)	1979
5	920,000	Wrexham, Wales (5.1)	1990
6	29,000,000	San Fernando (CA), USA (6.5)	1971
		northern Armenia (6.8)	1988
7	>890,000,000	Loma Prieta (CA), USA (7.1)	1989
		Kobe, Japan (7.2)	1995
		Izmit, Turkey (7.4)	1999
		Taichung and Nantou counties, Taiwan (7.6)	1999
		Rasht, Iran (7.7)	1990
		San Francisco (CA), USA (7.7–7.9)[1]	1906
8	>28,000,000,000	Tangshan, China (8.0)	1976
		Gansu, China (8.6)	1920
		Lisbon, Portugal (8.7)	1755
9	850,000,000,000	Prince William Sound (AK), USA (9.2)	1964

[1] Richter's original estimate of a magnitude of 8.3 was revised by two studies carried out by the California Institute of Technology and the US Geological Survey.

Appendix

Metamorphic Rocks

Typical depth and temperature	Main primary material (before metamorphism)		
	Shale with several minerals	Sandstone with only quartz	Limestone with only calcite
50,000 ft/570°F	slate	quartzite	marble
65,000 ft/750°F	schist		
82,000 ft/930°F	gneiss		
98,500 ft/1,100°F	hornfels	quartzite	marble

Mohs Scale

Number	Defining mineral	Other substances compared
1	talc	
2	gypsum	2½ fingernail
3	calcite	3½ copper coin
4	fluorite	
5	apatite	5½ steel blade
6	orthoclase	5¾ glass
7	quartz	7 steel file
8	topaz	
9	corundum	
10	diamond	

Note that the scale is not regular; diamond, at number 10 the hardest natural substance, is 90 times harder in absolute terms than corundum, number 9

Oil Spills

Year	Place	Source	Quantity	
			tonnes/tons	litres/gallons
1967	off Cornwall, England	Torrey Canyon	107,100	105,409
1968	off South Africa	World Glory	51,194,000	11,262,680
1972	Gulf of Oman	Sea Star	103,500	101,865
1977	North Sea	Ekofisk oilfield	31,040,000	6,828,800
1978	off France	Amoco Cadiz	200,000	196,842
1979	Gulf of Mexico	Ixtoc 1 oil well	535,000	526,553
1979	off Trinidad and Tobago	collision of Atlantic Empress and Aegean Captain	270,000	265,737
1983	Persian Gulf	Nowruz oilfield	540,000	531,474
1983	off South Africa	Castillo de Beliver	225,000	221,447
1989	off Alaska	Exxon Valdez	40,504,000	8,910,880
1991	Persian Gulf	oil wells in Kuwait and Iraq	6–8 million barrels	
1993	Shetland Islands, Scotland	Braer	85,000	83,657
1994	Komi province, northern Russia	ruptured pipeline	4.3–84.6 m	19.5–384 m
1996	South Wales	Sea Empress	72,500/71340	

Appendix

Major Late 20th-Century Earthquakes

As of March 2000.

(N/A = not available.)

Date	Location	Magnitude (Richter scale)	Estimated number of deaths
10 October 1980	northern Algeria	7.7	3,000
23 November 1980	southern Italy	7.2	4,800
13 December 1982	northern Yemen	6.0	1,600
30 October 1983	eastern Turkey	6.9	1,300
19, 21 September 1985	Mexico City, Mexico	8.1	5,000[1]
20 August 1988	Nepal/India	6.9	1,000
6 November 1988	southwestern China	7.6	1,000
7 December 1988	Armenia, USSR	6.8	25,000
17 October 1989	San Francisco (CA), USA	7.1	62
20–21 June 1990	northwestern Iran	7.7	50,000
16 July 1990	Luzon, Philippines	7.7	1,660
1 February 1991	Afghanistan/Pakistan	6.8	1,000
April 1991	northern Georgia	7.2	>100
20 October 1991	Uttar Pradesh, India	6.1	1,500
13, 15 March 1992	Erzincan, Turkey	6.7	2,000
12 December 1992	Flores Island, Indonesia	7.5	2,500
12 July 1993	western coast of Hokkaido, Japan	7.8	200
29 September 1993	Maharashtra, India	6.3	9,800
13–16 October 1993	Papua New Guinea	6.8	>60
6 June 1994	Cauca, Colombia	6.8	1,000

Appendix

Date	Location	Magnitude (Richter scale)	Estimated number of deaths
19 August 1994	northern Algeria	5.6	200
16 January 1995	Kobe, Japan	7.2	5,500
14 June 1995	Sakhalin Island, Russia	7.6	2,000
2 October 1995	southwestern Turkey	6.0	84
7 October 1995	Sumatra, Indonesia	7.0	>70
9 October 1995	Mexico	7.6	>66
3 February 1996	Yunnan Province, China	7.0	>250
17 February 1996	Irian Jaya, Indonesia	7.5	108
28 March 1996	Ecuador	5.7	21
4, 28 February 1997	Ardabil, Iran	N/A	>1,000
28 February 1997	Baluchistan Province, Pakistan	7.3	>100
10 May 1997	northeastern Iran (Khorasah Province)	7.1	>1,600
22 May 1997	India	6.0	>40
26 September 1997	central Italy	5.8	>11
28 September 1997	Sulawesi, Indonesia	6.0	>20
14 October 1997	north of Santiago, Chile	6.8	>10
21 November 1997	Chittagong, Bangladesh	6.0	17
11 January 1998	northeastern China	6.2	>47
4 February 1998	Takhar province, Afghanistan	6.1	>3,800
30 May 1998	northeast Afghanistan	6.9	4,000
27 June 1998	southern Turkey	6.2	145

Appendix

Date	Location	Magnitude (Richter scale)	Estimated number of deaths
19 November 1998	southwestern China	5.6 and 6.2	5
29 November 1998	Molluccas, Indonesia	6.5	>50
25 January 1999	western Colombia	6.0	900
29 March 1999	India	6.8	110
17 August 1999	Turkey	7.4	14,095
21 September 1999	Taiwan	7.6	2,256
30 September 1999	Mexico	7.5	70
12 November 1999	Turkey	7.2	737
27 November 1999	Vanuatu	7.2	10

[1] Some estimates put the death toll as high as 20,000.

Appendix

Major Floods and Tsunamis[1] Since 1981

[1] A tsunami is an ocean wave generated by vertical movements of the sea floor resulting from earthquakes or volcanic activity. As of January 2000.

Year	Event	Location	Number of deaths
1981	floods	northern China	550
1982	floods	Peru	600
	floods	Guangdong, China	430
	floods	El Salvador/Guatemala	>1,300
1983	tsunami	Japan/South Korea	107
1984	floods	South Korea	>200
1987	floods	northern Bangladesh	>1,000
1988	floods	Brazil	289
	floods	Bangladesh	>1,300
1990	tsunami	Bangladesh	370
1991	floods	Mexico	85
	floods	Tanzania	283
	floods	Afghanistan	1,367
	floods	Bangladesh	150,450
	floods	Benin	30
	floods	Chad	39
	floods/storm	Chile	199
	floods	China	6,728
	floods	India	2,024
	floods	Malawi	1,172
	floods	Peru	40
	floods/typhoon	Philippines	8,890
	floods	Romania	138
	floods	South Korea	54
	floods	Sudan	2,000
	floods	Turkey	30

Appendix

Year	Event	Location	Number of deaths
	floods	Texas, USA	33
	floods	Vietnam	136
1992	floods	Afghanistan	450
	floods	Argentina	104
	floods	Chile	41
	floods	China	197
	floods	India	551
	floods	Pakistan	1,446
	floods	Vietnam	55
1993	floods	Indonesia	18
	floods	midwestern USA	48
1994	floods	Moldova	47
	floods	southern China	1,400
	floods	India	>600
	floods	Vietnam	>175
	floods	northern Italy	>60
1995	floods	Benin	10
	floods	Bangladesh	>200
	floods	Somalia	20
	floods	northwestern Europe	40
	floods	Hunan Province, China	1,200
	floods	southwestern Morocco	136
	floods	Pakistan	>120
	floods	South Africa	147
	floods	Vietnam	85
1996	floods	southern and western India	>300
	floods	Tuscany, Italy	30
	floods	North and South Korea	86
	floods	Pyrenees, France/Spain	84
	floods	Yemen	324
	floods	central and southern China	2,300

Appendix

Year	Event	Location	Number of deaths
1997	floods	west coast, USA	36
	floods	Sikkim, India	>50
	floods	Germany/Poland/Czech Republic	>100
	floods	Somalia	>1,700
	floods	eastern Uganda	>30
	floods	Spain and Portugal	70
1998	tsunami	Papua New Guinea	>1,700
	floods/ mudslides	southern Italy	118
	floods	western Pakistan	>300
	floods	western Ukraine	17
	floods	Bangladesh	>400
	floods	northern India	>1,300
	floods	central, southeast, north, and northeast China	>3,650
	floods	Nepal	>250
	floods	South Korea	>195
	floods	Slovak Republic/Poland	>34
	floods	Kyrgyzstan/Uzbekistan border	>200
1998–99 (25 December– 6 January)	floods	Sri Lanka	>5 (approximately 155,000 affected and 15,000 families initially displaced)
1999	floods	Chang Jiang River, China	>925
	floods	Bangladesh	30
	floods	Nepal	97
	floods	Mexico	341
	floods	Thua Thien-Hua, Vietnam	470
	floods	southern France	31
	floods	Venezuela	30–50,000
2000	floods	Mozambique	>500

Appendix

Major Volcanic Eruptions in the 20th Century

Volcano	Location	Year	Estimated number of deaths
Santa María	Guatemala	1902	1,000
Pelée	Martinique	1902	28,000
Taal	Philippines	1911	1,400
Kelut	Java, Indonesia	1919	5,500
Vulcan	Papua New Guinea	1937	500
Lamington	Papua New Guinea	1951	3,000
St Helens	USA	1980	57
El Chichon	Mexico	1982	1,880
Nevado del Ruiz	Colombia	1985	23,000
Lake Nyos	Cameroon	1986	1,700
Pinatubo	Luzon, Philippines	1991	639
Unzen	Japan	1991	39
Mayon	Philippines	1993	70
Loki[1]	Iceland	1996	0
Soufrière	Montserrat	1997	23
Merapi	Java, Indonesia	1998	38

[1] The eruption caused severe flooding, and melted enough ice to create a huge sub-glacial lake.

Weathering

Physical weathering

Temperature changes	**Weakening rocks by expansion and contraction**
frost	wedging rocks apart by the expansion of water on freezing
unloading	the loosening of rock layers by release of pressure after the erosion and removal of those layers above

Chemical weathering

Carbonation	**Breakdown of calcite by reaction with carbonic acid in rainwater**
hydrolysis	breakdown of feldspar into china clay by reaction with carbonic acid in rainwater
oxidation	breakdown of iron-rich minerals due to rusting
hydration	expansion of certain minerals due to the uptake of water

Index

Page numbers in italics refer to illustrations (page number repeated where illustration and text on same page).
Useful terms are included in the A–Z glossary.

AAPG (American Association of Petroleum Geologists), 81
ABE (Autonomous Benthic Explorer) robotic submersible, 52
Absolute abundances of chemical elements, 37
Accelerators, 13
Acid rain, 49, 53
Adams, Leason Heberling, scientific contributions, 57
Adventures in the Learning Web, Web site, 99
AEG (Association of Engineering Geologists), 83
Aerosol cans, 47, 48
The Age of the Earth (Brown *et al*), 92
AGI (American Geological Institute), 81
AGU (American Geophysical Union), 81–2
AIH (American Institute of Hydrology), 82
Air pollution, 36, 40, 51
Aldrich, Lyman T, 38
Alexander, D *Natural Disasters*, 91
Altapedia Online Web site, 100
Alternative Technology Web site, 99
Alvarez, Luis W, 18, 49
ALVIN submersible, 19, 48 see also *A Water Baby* (Kaharl)
American Association of Petroleum Geologists (AAPG), 81
American Geological Institute (AGI), 81
American Geophysical Union (AGU), 81–2
American Institute of Hydrology (AIH), 82
American Institute of Professional Geologists, 82
Amino acids
  forms of, 17
  in rocks, 46
  synthetic, 17

Anderson, Don Lynn
  scientific contributions, 22, 57
  Theory of the Earth, 91
Annealing, optical glass, 57
Anschütz, Richard, 31
Antarctic
  conservation, 49
  research, 40, 45, 54, 55
Antarctica, Vostok station, 50
Antarctic Treaty, 42
Anthes, Richard Allen, scientific contributions, 57
Apocalypse (McGuire), 95
Aquifers Web site, 105
Arctic Circle Web site, 99
Arctic research, 56
Arctic Research Lab Ice Station II, 42
Argus experiment, 41
Arrhenius, Svante, 13
Artificial rain, 38, 40
Ask-A-Geologist Web site, 99
Ask an Energy Expert Web site, 99
Association for Exploration Geochemists, 83
Association for Women Geoscientists, 83
Association of Engineering Geologists (AEG), 83
Asteroid crashes
  effect on Earth, 49
  and extinction theory, 18
  on the Moon, 16
Asteroids, and early life, 17
Asthenosphere, 23
Astrogeology, 27
Astronomy, reading list, 94
Atlantic current theory, 74
Atmosphere
  composition, 229
  computer modelling, 57
  gases, 50–62

243

Index

greenhouse effect, 13, 45, 51, 56
layers, 31, 50
mathematical model, 47, 50
measuring distance above Earth, 35
oxygen abundance, 16
and pollution, 13–14, 51
and radio waves, 31
reading list, 91
research programmes, 38, 41, 47, 49, 50, 62
temperature, 31, 34
volcanic activity and, 16
wind research, 37
see also Ozone layer
Atmosphere, Weather and Climate (Barry and Chorley), 91
Atmospheric pressure, instability theory, 61
Atomic bomb, development, 75
Atomic bonding, 4
Atomic research, 10
Audubon First Field Guide: Rocks and Minerals (Ricciutti and Carruthers), 96
Autonomous Benthic Explorer (ABE) robotic submersible, 52
Avalanche Awareness Web site, 100
Avalanche! Web site, 100

Bailey, Edward Batersby, scientific contributions, 58
Barghorn, Elso Sterrenberg, 46
Barnes, J W *Basic Geological Mapping*, 91
Barringer crater, 27
Barry, Roger G *Atmosphere, Weather and Climate*, 91
Barton, Otis, 36
Basic Geological Mapping (Barnes), 91
Basic Palaeontology (Benton), 91
Basins, ocean, 22, 45
Bathyscaphs, 39, 42
Bathyspheres, 36, *36*
Batson, R *NASA Atlas of the Solar System*, 94
Bavarian Research Institute of Experimental Geochemistry and Geophysics, 9
Beebe, William, 36
Before and After the Great Earthquake and Fire: Early Films of San Francisco, 1897–1916 Web site, 100

Bell, Peter, 9, 47
Benioff, Victor Hugo, scientific contributions, 38, 58
Benioff-Wadati zones, 38, 58
Benton, MJ *Basic Palaeontology*, 91
Big Empty Web site, 100
Biodiversity, 52, 54
reading list, 98
Biogeochemistry, 12, 17
Biology
brain evolution, 63
gases from micro-organisms, 62
marine biology, 49
reading list, 93, 97
undersea life forms, 48
Biosphere, research programmes, 49, 51
Birch, (Albert) Francis, scientific contributions, 42, 58
Bird, J M, 46
Bjerknes, Jacob, 33, 39
Bjerknes, Vilhelm Firman Koren, scientific contributions, 33, 58–9
Blue Planet: An Introduction to Earth System Science (Skinner), 96
Bodleian Library Map Room – The Map Case Web site, 101
Bolt, BA *Earthquakes*, 91
Boltwood, Bertram, scientific contributions, 10, 59
Bolzmann, Ludwig, 2–3
Bonds (chemistry), 4–5, 13
The Book of Life (Gould ed), 94
Bort, Léon-Philippe Teisserenc de, 31
Bottomley, Richard, 55
Bowen, Norman Levi, scientific contributions, 4, 35, 59
Bradley, Albert, 52
Bragg, Lawrence, 37
Brain, evolution, 63
Bridgman, Percy Williams, scientific contributions, 7, 32, 60
Briggs, David *Encyclopedia of Earth Sciences*, 91
British Geological Survey Global Seismology and Geomagnetism Group Earthquake Page Web site, 101
British Geological Survey Web site, 101
British Petroleum, *45*

Index

Broecker, W *How to Build a Habitable Planet*, 91
Brown, G A *The Age of the Earth*, 92
Bullard, Edward Crisp, scientific contributions, 38, 60, *61*
Byrd Station research, 45

Calder, Nigel
   Restless Earth, 92
   The Weather Machine, 92
Cambrian Explosion, 55
Cambrian Period: Life Goes For a Spin Web site, 101
The Cambridge Encyclopedia of Earth Sciences (Dixon), 93
Carbon dating, 70
Carbon dioxide, atmospheric, 13–14
Carnegie Institution of Washington DC, 4, 9
Carnegie survey vessel, *34*, 35
Carruthers, M
   Audubon First Field Guide: Rocks and Minerals, 96
   Pioneers of Geology: Discovering Earth's Secrets, 92
Carson, Rachel, 43
Cartographic Images Home Page Web site, 101
Cas, R A F *Pioneers of Geology: Discovering Earth's Secrets*, 92
Catastrophism, 28
Cauldron subsidences, 58
Causes of Climate Change Web site, 102
CFCs (Chlorofluorocarbons), 47, 48, 50–51, 71
Charney, Julie Gregory, scientific contributions, 61–2
Chemical equilibrium, laws, 67
Chemical thermodynamics, 2–4
Chemistry
   absolute abundances of chemical elements, 37
   atmospheric, 62
   biogeochemistry, 12, 17
   bonds, 4–5, 13
   chemical thermodynamics, 2–4
   of Earth's near surface, 11
   environmental surface chemistry, 5
   geochemistry, 17
   and geology, 35
   high-pressure minerals, 4, 7, 39
   mineral chemistry, 4–5, 9
   of ozone layer, 62
   pesticides, 43
   see also Dating methods; Geochemistry; Isotopes
Chicxulub crater, 18, 52, 54
Chlorofluorocarbons (CFCs), 47, 48, 50–51, 71
Chorley, Richard J *Atmosphere, Weather and Climate*, 91
Chromatography, 17
Claude, Georges, 35
Clean Air Act, 40
Cleve, Per Teodor, 31
Climate, 13–15
   classification, 31
   climatic fluctuations, 13–15
   conferences, 53
   Earth summits, 54
   patterns, 61–2
   reading list, 91, 93, 95
   research programmes, 48, 52
   temperature, 13
   Web sites, 101, 103, 104, 108, 116, 118
   see also Meteorology; Weather
Climate and the British Scene (Manley), 95
Climatology, 75
Clinton, S *Pioneers of Geology: Discovering Earth's Secrets*, 92
Cloos, Ernst, scientific contributions, 62
Clouds and Precipitation Web site, 102
Cloud-seeding, 38, 40
Clouds from Space Web site, 102
Coastal Features and Processes Web site, 102
Coasts in Crisis Web site, 102
Coesite, 7
Coes, Loring, 7, 39
Collar, 7, 32
Comet crashes, 18, 49
Communications
   disruption by magnetic storms, 47
   Telstar, 40
Composition of Rocks Web site, 102
Computational mineral physics, 4
Computers, Atlas, computer, 43
Coney, P, 49

Index

Conservation
  Antarctic conservation, 49
  international conferences, 50
  organizations, 55
  pollution legislation, 48
  see also Environmental sciences
Conshelf Saturation Dive Programme, 44
Continental displacement theory, 20
Continental drift, 35, 44, 60, 62, 68, 77–8
  theory, 20, 63, 69
  see also Plate tectonics; Seafloor spreading
Continent formation, 23, 63
Continents, evolution, reading list, 98
Convention on Conservation of Antarctic Marine Living Organisms, 49
Core of Earth, 6, 9, 10, 54
  discovery, 74
  electric currents, 64
  refraction of seismic waves, 74
  see also Deep interior of Earth
Coriolis Effect Web site, 103
Cosmic abundance of elements, 67
Cosmic radiation, 32
Cosmogenic nuclides, 13
Cousteau, Jacques, 44
Cowling, George, 39
Cox, Allan V(erne), scientific contributions, 44, 62
Cracking the Ice Age Web site, 103
Craters
  Barringer crater, 27
  Chicxulub crater, 18, 52, 54
  Popigai crater, 55
Crust of Earth, 63, 65, 66, 67
Crutzen, Paul, scientific contributions, 62
Crystallography, 67
Crystal structures, prediction of, 35

Dalrymple, Brent, 44
Dan's Wild Wild Weather Page Web site, 103
Dana, J D *Dana's New Mineralogy: The System of Mineralogy of James Dwight Dana and Edward Salisbury Dana*, 92
Dasch, E Julius *Sedimentary Environments: Processes, Facies and Stratigraphy*, 92

Dating methods
  carbon dating, 70
  cosmogenic nuclide decay, 13
  isochron method, 42
  isotope dating, 10–11, 17, 33, 40
  magnetic polarity reversal, 44
  potassium-argon dating, 38
  radioactive dating, 33, 59, 68
  radiocarbon dating, 70
  rubidium-strontium dating, 42
  uranium-lead dating, 42, 59, 69
de Bort, Léon Teisserenc -Philippe, 31
de Geer, Gerhard, 32
de la Mare, William, 55
Decker, R W and Decker, B B *Mountains of Fire: The Nature of Volcanoes*, 92
Deep interior of Earth, 4, 5–10 *see also* Core of Earth
Deep Sea Drilling Project (DSDP), 44, 47, 49, 50
Deer, W A *Rock-forming Minerals*, 92
Department of Atmospheric Sciences Web site, 103
Detrich, Bill, 54
Deuterium, (heavy hydrogen), 11, 75
Dewey, John F, 46
Diamond anvil cell, 8–9, *8*
Diamonds, synthetic, 7, 60
Differentiation, 3, 4, 35
Dinofish.com Web site, 103
Dinosaurs, 12
  extinction theories, 18, 49
  reading list, 95
Disasters (Whittow), 98
Disasters: Panoramic Photographs, 1851–1991 Web site, 103
Discoverer 1 satellite, 41
Discovering Landscape (Goudie and Gardner), 93
The Diversity of Life (Wilson), 98
Diving programmes, 44
Dixon, Dougal *The Cambridge Encyclopedia of Earth Sciences*, 93
Doell, Richard, 44
Donald L Blanchard's Earth Sciences' Web site, 104
Doppler radar, 42
Double Whammy Web site, 104

Index

Drilling
  drilling projects, 44, 47–50
  ice drilling, 52, 54
  ocean drilling, 44, 46, 47–50
  record-breaking, 49, 52
Drury, S A *Image Interpretation in Geology*, 93
DSDP (Deep Sea Drilling Project), 44, 47, 49, 50
Du Toit, Alexander Logie, scientific contributions, 63
Duff, McL D *Holmes' Principles of Physical Geology*, 93
Dunning, F W *The Story of the Earth*, 93
Dust blown from Sahara, 54
Dust Bowl Web site, 104
Dynamo model of Earth, 37, 60, 64
Dziewonski, Adam Marian, scientific contributions, 6, 63

Earth and Heavens – The Art of the Map Maker Web site, 104
Earth and Life Through Time (Stanley), 97
Earth Introduction Web site, 104
Earth sciences, 1–2
Earth Sciences Web site, 105
Earth summits, 54
Earth
  atmosphere, 31
  boundary between crust and mantle, 32, 72
  boundary between lower mantle and outer core, 33
  constituents of, 37, 49, 58
  core, 6, 9, 10, 54, 64, 74
  crust, 63, 65, 66, 67
  dating, 10–11, 33, 40, 59, 68
  deep interior, 4, 5–10
  depletion of resources, 46
  dynamo model, 37, 60, 64
  evolution, 57
  formation theory, 76
  Gaia hypothesis, 71
  gravitational field, 49
  heat-flow through ocean floor, 38, 60
  inner core, 37, 68
  layers of, 6–7, 20–1
  life formation experiment, 72
  magma, 59
  magnetic field, 5, 21, 41, 43, 49, 66, 71
  mantle, 6, 35, 39, 42, 47, 68
  near surface chemistry, 11
  orbit and temperatures, 15
  pear-shaped, 41
  radiation belt, 41
  reading list, 91, 92, 93, 95, 96
  reversal of magnetic field, 35, 62
  space research, 41, 47, 48, 52
  Web sites, 99, 100, 104, 106, 108, 117
  see also Atmosphere; Climate; Deep interior of Earth; Geology; Oceanography
Earth's Seasons: Equinoxes, Solstices, Perihelion and Aphelion Web site, 105
Earthquakes
  faults, 38
  late 20th century, 234–6
  magnitude, 68
  nuclear detection system, 58
  prediction, 23, 24, 48
  reading list, 91, 93, 97
  record-breaking, 37
  San Francisco (1906), 24
  seismic waves, 74
  Web sites, 100, 104, 105, 112, 122
  see also Seismology
Earthquakes (Bolt), 91
Earthquakes (van Rose), 97
Earthquakes and Plate Tectonics Web site, 104
Edinger, Tilly, scientific contributions, 63, 64
Edwards Aquifer Homepage Web site, 105
El Niño, 49, 55
El Niño Theme Page Web site, 105
El Niño Web site, 105
Elasticity
  of minerals, 57, 58
  of rocks, 57, 58
Electric currents in Earth's core, 64
Electricity generation, ocean temperatures demonstration, 35
Electromagnetic frequencies, Earth as conductor of, 31
Electron capture detector, 71
Electronegativity scale, 4
Elements, cosmic abundance, 67
Elsasser, Walter Maurice, 37, 64
Elsom, D *Planet Earth: the Making, Shaping and Workings of a Planet*, 93

Index

Emery, Dominic *Planet Earth: the Making, Shaping and Workings of a Planet*, 93
Encyclopedia of Earth Sciences (Briggs), 91
Encyclopedia of Volcanoes (Sigurdsson), 96
Environment *see* Conservation; Environmental sciences; Pollution
Environmental sciences
  environmental surface chemistry, 5
  ozone layer damage, 47, 48, 62
  Web sites, 100, 110, 119, 121
  see also Conservation
Eocene geological period, 55
EqIIs – Earthquake Image Information System Web site, 105
Equatorial Undercurrent, 40
Erosion and Deposition Web site, 105
ERS satellites, 52, 53
Eskola, Pentti Eelis, scientific contributions, 3, 33, 65
Essential Guide to Rocks Web site, 106
Essentials of Oceanography (Thurman), 97
European Space Agency, 52, 53
Evolution
  brain, 63
  Earth, 57
  fossil research, 16, 50, 63, 67
  of plates, 57
  reading list, 97
The Evolution of the Igneous Rocks, (Bowen), 4
Evolving Continents (Windley), 98
Ewing, (William) Maurice, scientific contributions, 39, 40, 65
Exploring the Environment Web site, 106
Exploring the Tropics Web site, 106
Explosion seismology, 35
Extinction
  mass extinction, 18
  surveys, 55
  theories, 18, 49, 55
Exxon Valdez oil tanker, 51

Fabry, Charles, 33
Facies, metamorphic, 33
Facing the Future: People and the Planet Web site, 106
FAMOUS (French-American Mid-Ocean Undersea Study), 19, 48

Faults, 26, 38, 45
  fault mechanics, 24
  transform faults, 22, 69
Features Produced by Running Water Web site, 106
Fernando de Noronha Web site, 106
Field Palaeontology (Goldring), 93
Fleming, John Adam, scientific contributions, 65–6
Floods
  reading list, 91
  since 1981, 237–9
Flood! Web site, 107
Fortey, Richard *Fossils: The Key to the Past*, 93
Fossils
  amphibian, 50
  Cambrian, 32, 55
  classification, 74
  dating, 17, 44, 69, 70
  Devonian, 66
  evolutionary research, 16, 50, 63, 67
  microscopic organisms, 39
  multicellular animals, 16, 51
  palaeoneurology, 63
  pre-metazoan, 16, 39
  reading list, 91, 93, 96, 97
  see also Palaeontology
Fossils (Swinnerton), 97
Fossils: The Key to the Past (Fortey), 93
Francis, P *Volcanoes: A Planetary Perspective*, 93
French-American Mid-Ocean Undersea Study, (FAMOUS), 19, 48
Front (origin of term), 33, 59

The Gaia Atlas of Planet Management (Myers), 95
Gaia hypothesis, 71
Gardner, Julia (Anna), scientific contributions, 66
Gardner, Rita *Discovering Landscape*, 93
GARP (Global Atmospheric Research Program), 47, 50
Gases
  atmospheric, 50, 62
  electron capture detector, 71
  greenhouse gases, 13–14, 56
  from micro-organisms, 62
  'primordial soup', 17
  from undersea hot springs, 19

Gass, Ian G, 43
Geer, Gerhard de, 32
Genetic modification (GM), 53
Geochemical Society, 83
Geochemistry, 17, 35
    Earth's crust elements, 66
    isotope geochemistry, 38
    organizations, 83
    structural change of minerals, 44
    see also Isotopes
Geodesy, reading list, 96
Geodynamics, satellite research, 44
Geographia Web site, 107
Geographical Study Resources Web site, 107
Geological Association of Canada, 84
Geological Society of America (GSA), 84
Geological Society (UK), 84
Geological time, 10–11
Geologists' Association, 85
Geology, 1
    astrogeology, 27
    catastrophism, 28
    dating methods, 42
    Eocene period, 55
    K–T boundary, 18
    mapping, 66
    magnetism of rocks, 5, 35, 43, 55, 72
    military, 66
    of the Moon, 16
    ocean floor, 65, 68
    Oligocene period, 55
    ophiolites, 43
    organizations, 81, 82, 83, 84, 85, 88
    overthrusting, 41
    reading list, 92, 93, 94, 96, 97, 98
    of the Solar System, 16, 27–9
    timescale, 62
    uniformitarianism, 28
    Web sites, 99, 100, 101, 102, 107, 119, 120
    see also Geochemistry; Geophysics; Minerals; Mineralogy; Petrology; Rocks
Geology and Scenery in England and Wales (Trueman), 97
Geology Link Web site, 107
Geomagnetism, 63, 65, 66
Geomorphology, reading list, 97

Geomorphology and Global Tectonics (Summerfield), 97
Geophysical Laboratory, Carnegie Institution of Washington DC, 4
Geophysics
    Carnegie survey vessel, 34, 35
    heat-flow measurement, 38, 60
    high-pressure physics, 7–9, 32, 47, 60
    magnetism of rocks, 5, 35, 43, 55, 72
    marine, 60
    organizations, 81–2, 86
    prospecting, 5, 34
    rifts, 39
Geoscience, organizations, 83, 85
Geoscience Information Society, 85
Geothermometers, isotopes as, 11, 38
GEOS satellite, 44
Gibbs, J Willard, 2–3
Gilbert, Grove Karl, 27
Glacier Web site, 107
Glaciers Web site, 108
Glaucophane, 69
Global Atmospheric Research Program (GARP), 47, 50
Global Climate Change Information Programme (GCCIP) Web site, 108
Global Drainage Basins Database Web site, 108
Global Geomorphology: an Introduction to the Study of Landforms (Summerfield), 97
Global Positioning System (GPS), 23
Global seismic tomography, 6
Global Tectonics (Kearey and Vine), 95
Global warming, 13–15, 52, 53, 56
    Kyoto Protocol, 56
    reading list, 94
Global Warming: The Complete Briefing (Houghton), 94
Glomar Challenger research vehicle, 48, 49, 50
Glossary of Geology (Jackson), 94
GM (genetic modification), 53
GOES (Geostationary Operational Environmental Satellite), 47
Goldring, R Field Palaeontology, 93
Goldring, Winifred, scientific contributions, 66
Goldschmidt, Victor Moritz, 3, 35, 37, 66–7
Gondwanaland, 63

Index

Goudie, Andrew *Discovering Landscape*, 93
Gould, Stephen Jay
  (ed) *The Book of Life*, 94
  scientific contributions, 67
GPS (Global Positioning System), 23
Gravimetry, 72
Gravitation fields
  Earth's changing, 49
  measurement device, 34
  see also Torsion balance
Gravity
  determination techniques, 72
  measuring variations, 60
Greatest Places Web site, 108
Great Globe Gallery Web site, 108
Greeley, R
  NASA Atlas of the Solar System, 94
  Planetary Landscapes, 94
Greenhouse effect, 13, 45, 51, 56
  greenhouse gases, 13–14, 56
  Web site, 117
The Greenhouse Effect: How the Earth Stays Warm Web site, 117
The Greenpeace Book of Coral Reefs (Wells and Hanna), 98
Gribbin, John *The Weather Book*, 94
Ground Beneath Web site, 108
Groundwater Quality and the Use of Lawn and Garden Chemicals by Homeowners Web site, 109
GSA (Geological Society of America), 84
Gulf Stream dynamics, 74
Gunflint biota, 16
Gutenberg, Beno, scientific contributions, 6, 33, 37, 68
Gyrocompass, 32

Hall, A *Igneous Petrology*, 94
Hallam, A *A Revolution in the Earth Sciences*, 94
Hanna, Nick *The Greenpeace Book of Coral Reefs*, 98
Hardy, Ralph *The Weather Book*, 94
Hayes sonic depth finder, 33
Heat-flow measurement, 38, 60
Heavy hydrogen (deuterium), 11, 75
Heavy water, 75
Heezen, Bruce, 40

Hess, Harry Hammond, scientific contributions, 42, 43, 68
Hess, Victor Francis, 32
High-pressure minerals, 4, 7, 39
High-pressure physics, 7–9, 32, 47, 60
High-temperature physics, 60
History of Ocean Basins (Hess), 22
Hoering, T C, 17
Holmes, Arthur, scientific contributions, 20, 33, 35, 68–9
Holmes' Principles of Physical Geology (Duff), 93
Hot ice, 7
Houghton, J *Global Warming: The Complete Briefing*, 94
How to Build a Habitable Planet (Broecker), 91
Hubbert, Marion, 41
Hurricane and Storm Tracking Web site, 109
Hurricane-seeding, 38, 43
Hutton, James, 28
Hydrodynamics, 75
Hydrologic Cycle Web site, 109
Hydrology, 58
  organizations, 82
Hydrology Primer Web site, 109

Ice research, 14–15, 52, 54
Ice stations, 42, 45, 50, 54
Igneous Petrogenesis, a Global Tectonic Approach (Wilson), 98
Igneous Petrology (Hall), 94
Igneous rocks, 35, 58
Image Interpretation in Geology (Drury), 93
Impact craters, 27–9
  Chicxulub, 18, 52, 54
  Popigai, 55
Impact sites, K-T, 52
Inner core of Earth, 37, 68
Instability theory of atmospheric pressure, 61
Internal and External Friction Web site, 109
International Magnetosphere study, 47
International Phase of Ocean Drilling (IPOD), 47
International Union for the Conservation of Nature (IUCN), 55

Index

Introduction to Environmental Education Web site, 110
Introduction to Geodesy: the History and Concepts of Modern Geodesy (Smith), 96
Introduction to Plate Tectonics Web site, 110
Introduction to the Ecosystem Concept Web site, 110
Ionizing layer of atmosphere, 35
IPOD (International Phase of Ocean Drilling), 47
Isochron dating method, 42
Isomagnetic world charts, 66
Isotopes
  carbon isotopes, 17
  cosmogenic nuclides, 13
  dating methods using, 10–11, 17, 33, 40
  discovery of, 10, 59
  heavy isotopes, 11, 50, 76
  isotope geochemistry, 38
  oxygen isotopes, 12–13
  stable isotopes, 11, 17, 38, 67
  uranium, 70
Ittekkot, Venugopalan, 54
IUCN (International Union for the Conservation of Nature), 55

Jackson, J *Glossary of Geology*, 94
JOIDES (Joint Oceanographic Institutions for Deep Earth Sampling), 44, 46, 50
Jones, D L, 49

K–T boundary, 18
K–T impact site, 18, 52
Kaharl, Victoria A *A Water Baby*, 94
Kearey, P *Global Tectonics*, 95
Khark 5 oil tanker, 51
Kington, John *The Weather Book*, 94
Kirschvink, Joseph L, 55
Klein, C *Manual of Mineralogy (after James D Dana)*, 95
Knopf, Leonora Frances, scientific contributions, 69
Köppen, Vladimir Peter, 31
Korzhinskii, D S, 3, 37, 41
Kyoto Protocol, global warming, 56

Lageos (Laser Geodynamic satellite), 48, 49

Landforms of Weathering Web site, 110
Landsat satellites, 46, 47
Langmuir, Irving, 38
Large Aperture Seismic Array, 44
Lasers, 9, 11, 48, 26
Late Pleistocene Extinctions Web site, 110
Laurasia, 63
Le Pichon, Xavier, scientific contributions, 22, 46, 69
Lehmann, Inge, 6, 37
Libby, Willard Frank, scientific contributions, 70, *70*
Life on Earth, 15–19
  damage to sealife, 54
  discovery of new life forms, 19
  recreation experiment, 72
  undersea life forms, 48
Light, research, 25
Lindemann, Frederick, 34
Lithosphere, plates of, 22, 38, 44, 46, 69
Liu, John, 9, 47
Lovelock, James Ephraim, scientific contributions, 71, *71*
Lunar and Planetary Institute, 85
Lunar exploration, reading list, 98
Lunar physics, 65
Lyell, Charles, 28

Magellan, Ferdinand 42
Magma, 59
Magnetic fields
  anomalies, 46, 76
  dynamo model, 37, 60, 64
  mapping, 49
  measurement, 5, 21
  polarity stripes, 22, 43
  reversal of, 21, 35, 44, 62, 71–2
  in rocks, 5, 35, 43, 55, 72
Magnetic storms, 47
Magnetosphere, 47
Magsat satellite, 49
Major, A, 44
Malloy, Kirk, 54
Manabe, Syukovo, 45
Manhattan Project, 70, 75
Manley, Gordon *Climate and the British Scene*, 95
Mantle of Earth, 6, 35, 39, 42, 47, 68
Manual of Mineralogy (after James D Dana) (Klein), 95

Index

Mao, David, 9, 47
Mapping
  Earth's magnetic field, 49
  geological, 66
  ocean floor mapping, 21
  reading list, 91
  remote sensing, 25–6
  Web sites, 100, 101, 104, 110, 111, 113, 117
Mapquest Web site, 110
Map Quiz Tutorial: Physical Geography Web site, 111
Marconi, Guglielmo, 31
Mare, William de la, 55
Mariana Trench, Pacific Ocean Web site, 111
Marine Geology and Geophysics Division of the NOAA National Geophysical Data Center, 85
Marine sciences *see* Oceanography
Mars, 18–19, 28
Mass extinction, 18
Mass spectrometers, *12*, 13
Mathematics of Cartography Web site, 111
Matsuyama Epoch, 72
Matsuyama, Motonori, scientific contributions, 35, 71–2
Matthews, Drummond, 22, 43
McGuire, B *Apocalypse*, 95
McKenzie, Dan, 22, 43
Meadows, Dennis, 46
The Meaning of Fossils (Rudwick), 96
Melting the Earth: the History of Ideas on Volcanic Eruption (Sigurdsson), 96
Mercalli scale, 230
The Message of Fossils (Tassy), 97
Metallurgy, organizations, 88
Metals
  high-pressure research, 8–9
  melting points, 10
Metamorphic facies, 33
Metamorphic rocks, 65, 69, 232
Meteorites, 13
  dating, 11, 40
  and Earth's atmosphere, 16
  impact craters, , 18, 27–9, 52, 54, 55
Meteorology, 57, 58
  cloud-seeding, 38, 40
  computers for, 38, 42, 61
  dust blown from Sahara, 54
  equipment, 37
  hurricane-seeding, 38
  first mathematical techniques, 34
  movement of air masses, 33
  photography, 39
  radar, 42
  reading list, 94
  remote sensing, 24–5, 26
  satellites, 43, 47, 50
  first TV weather man, 39
  weather stations, 37, 39
  Web sites, 102, 109, 111, 121
  wind research, 37
  see also Climate; Weather
Meteors, 27–9, 34
Met Office Homepage Web site, 111
Meyers, Keith *Planet Earth: the Making, Shaping and Workings of a Planet*, 93
Milankovitch cycles, 15, 33
Milankovitch, Milutin, 15, 33
Military Geologic Unit, 66
Miller, Stanley Lloyd, scientific contributions, 17, 72, *73*
Milner, Angela *The Natural History Museum Book of Dinosaurs*, 95
Mineralogical Society of America, 86
Mineralogy, 2–5
  chemistry, 4–5, 9
  Earth's mantle, 39
  of mountains, 3
  organizations, 86
  physics, 4, 9, 32
  reading list, 92, 95, 96
  structural change of minerals, 44
Mineralogy Database Web site, 111
Minerals
  dating, 10–11, 38
  elasticity, 57, 58
  high-pressure minerals, 4, 7, 39
  Mohs scale, 232
  reading list, 92,
  structures, 4, 44
  synthetic, 7
  Web site, 111
  see also Geology; Mineralogy; Petrology; Rocks
Miyashiro, Akilio, 38
Models of Landform Development Web site, 111

Index

Modified Mercalli Intensity Scale Web site, 112
Mohole Project, 68
Mohorovičić, Andrija, scientific contributions, 6, 32, 72–3
Mohorovičić discontinuity, 6, 32, 72
Mohs scale, 232
Molina, Mario, 47, 62
Monger, J W H, 49
Montreal Protocol, 50
Moon, 16, 18, 27–8
Morgan, Jason, 22, 44
Morley, L W, 22, 43
Moulin, Cyril, 54
Mountains
  formation of, 21, 23, 43, 49
  measuring, 26
  and mineralogy, 3
  undersea, 21
Mountains of Fire: The Nature of Volcanoes (Decker and Decker), 92
MTU Volcanoes Page Web site, 112
Multimedia History of Glacier Bay, Alaska Web site, 112
Myers, Norman *The Gaia Atlas of Planet Management*, 95

Naming of Atlantic Hurricanes Web site, 112
Nappes, 58
NASA Atlas of the Solar System (Greeley and Batson), 94
National Aeronautics and Space Administration (NASA), 43, 71 *see also* Space
National Conservation Congress, 32
National Geographic Online Web site, 112
National Geophysical Data Center, National Oceanic and Atmospheric Administration (NOAA), 54, 85, 86
National Oceanic and Atmospheric Administration (NOAA) Web site, 112
National Weather Service, National Oceanic and Atmospheric Administration (NOAA), 86
Natural disasters
  reading list, 95, 98
  refugees, 56
  Web sites, 100, 103, 107
  see also Floods; Tsunamis; Volcanoes

Natural Disasters (Alexander), 91
The Natural History Museum Book of Dinosaurs (Milner), 95
Nature Explorer Web site, 112
The Nature of the Chemical Bond (Pauling), 4
Neumann, John von, 38
Nicolaysen, L O, 42
Nier, Alfred, 38
NOAA (National Oceanic and Atmospheric Administration), 54, 85, 86
Nolet, G *Seismic Tomography*, 95
North American Cordilleran orogen, 49
Nuclear detection system, earthquakes, 58
Nuclear testing, 41

O'Neill, Alan, 55
Ocean Basins: Their Structure and Evolution (Open University Oceanography course team), 96
Ocean Drilling Program (ODP), 50
Ocean floor
  drilling, 44, 46, 47–50
  geology, 65, 68
  mapping, 21
  organizations, 44, 85
  surveys, 26
  see also Oceanography; Oceans; Seafloor spreading
OceanLink Web site, 113
Oceanography
  electricity generation, 35
  heat-flow through ocean floor, 38, 60
  marine biology, 48, 49
  marine geophysics, 60
  ocean basin formation, 45
  reading list, 94–5, 95–6, 97, 98
  research programmes, 52
  Web sites, 112, 113, 121
  see also Oceans; Ocean floor
Ocean Planet Web site, 113
Oceans
  basins, 22, 45
  damage to sealife, 54
  diving programs, 44
  exploration, 36

253

undersea hot springs, 19
undersea mountains, 21
see also Ocean floor; Oceanography
Ocean Satellite Image Comparison Web site, 113
Ocean Web sites, 111, 113
ODP (Ocean Drilling Program), 50
OECD (Organization for Economic Cooperation and Development, 53
Oil exploration, 5, 34, 35
Oil Pollution Act, 34
Oil spills, 45, 48, 51, 52, 233
Oldham, Richard Dixon, scientific contributions, 5, 31, 74
Oligocene geological period, 55
Open University Oceanography course team *Ocean Basins: Their Structure and Evolution*, 96
Ophiolites, 43
Orbit of Earth, relation to temperature, 15
Ordnance Survey Web site, 113
Organization for Economic Cooperation and Development (OECD), 53
Orogenic belts, 49, 69
Out of This World Exhibition Web site, 113
Overpopulation, 46, 49
Overthrusting, 41
Oxburgh, Ernest R, 46
Oxford University Oceanography course team *Waves, Tides and Shallow Water Processes*, 95
Oxygen
  abundance in Earth's atmosphere, 16
  isotopes, 12–13, 50
  and photosynthesis, 16
Ozone, oxygen isotopes in, 12
Ozone layer, 50
  chemistry, 62
  damage, 47, 48, 62
  depletion, 52, 54
  discovery, 33
  hole, 56
  increased UV radiation, 54
  measurement, 53

Pakicetus skull, 50
Palaeomagnetics, 62
Palaeoneurology, 63
Palaeontological Research Institution, 87

Palaeontology, 66
  organizations, 87
  punctuated equilibrium theory, 67
  reading list, 91, 93, 94
  stratigraphic, 66
  vertebrate, 63
  Web sites, 103, 104, 110
  see also Fossils
PALEOMAP project Web site, 113
Pangaea, 20, 32, 77
Paraphyletics, 74
Parker, R L, 44
Patterson, Colin, scientific contributions, 40, 74
Pauling, Linus, 4
Pelican Island wildlife refuge, 31
Pellant, Chris *The Practical Geologist*, 96
Pesticides, 43
Petrofabrics, 69
Petroleum
  exploration, 5
  research, 17
Petrology, 38, 59, 65, 69
  organizations, 88
  reading list, 94
  see also Geology; Rocks
Photosynthesis, 16
Physics
  high-pressure physics, 7–9, 32, 47, 60
  high-temperature physics, 60
  lunar, 65
  mineral physics, 4, 9, 32
  solar, 65
  Web sites, 103, 109, 116
  see also Geophysics; Thermodynamics
Piccard, Jacques, 42
Pidgeon, Bob, 50
Pierce, John, 40
Pioneers of Geology: Discovering Earth's Secrets (Carruthers and Clinton), 92
Planetary Landscapes (Greeley), 94
Planet Earth: the Making, Shaping and Workings of a Planet (Elsom), 93
Planet Earth: the Making, Shaping and Workings of a Planet (Emery and Meyers), 93
Planets
  composition of, 9–10
  geology, 27–9
  Mars, 18–19, 28

organizations researching, 9
Venus, 29
see also Space
Plates
  construction, 6, 22
  evolution, 57
  motions 22–3, 44, 46
  oceanic ridges, 40
  splitting, 53
Plates of lithosphere, 22, 38, 44, 46, 69
Plate tectonics, 2, 5, 20–3, 63, 77, 78
  continental drift, 35, 69
  ocean floor geology, 65, 68
  plate construction, 6, 22
  reading list, 93, 94, 95
  seafloor spreading, 44, 68, 76
  Web sites, 104, 108, 110, 114, 115, 117
  see also Continental drift; Seafloor spreading; Terrane concept
Plate Tectonics Web site, 114
Platt, J H, 59
Polar bears, 56
Polar front theory, 58–9
Polar, research, 14 *see also* Antarctic research; Arctic research
Pollution, 46
  air pollution, 36, 40, 51
  atmospheric pollution, 13–14, 51
  international agreements, 50–1
  legislation, 48
  oil pollution, 34, 45, 48
  ozone depletion, 50, 54
  policies, 40
  transport of hazardous waste, 53
  US reports on, 49
Popigai impact crater, 55
Population control, 46, 49
Potassium-argon dating, 38
Power plants, sulphur emissions, 49
The Practical Geologist (Pellant), 96
Precambrian rocks, 65
Press, Frank *Understanding Earth*, 96
Primary Geography Page Web site, 114
'Primordial soup', 17
Pulse modulation technique, 35
Punctuated equilibrium theory, 67

Quantum mechanics, 11
Questions and Answers About Snow Web site, 114

Radar, Doppler, 42
Rader's Terrarum Web site, 114
Radiation, cosmic, 32
Radioactivity
  discovery of, 10
  radioactive dating, 33, 59, 68
  in rocks, 20
Radiocarbon dating, 70
Radio meteorograph (radiosonde), 37
Radio waves, 31
Rain, artificial, 40
Rainforests, 52, 56
Rainforests of the World Web site, 114
Reading, Harold *Remote Sensing*, 96
Recapitulation, 67
Recumbent folds (nappes), 58
Remote sensing, 25–7
Remote Sensing (Reading), 96
Restless Earth (Calder), 92
A Revolution in the Earth Sciences (Hallam), 94
Ricciutti, E *Audubon First Field Guide: Rocks and Minerals*, 96
Richardson, Lewis Fry, 34
Richter, Charles Francis, scientific contributions, 37, 74
Richter scale, 74, 231
Rifts, 39
Ringwood, A E, 44
River Gauging Stations Web site, 115
Rivers and Streams Web site, 115
Rivers Web sites, 115, 116
Rock-forming Minerals (Deer), 92
Rocks
  amino acids in, 46
  dating, 11, 33, 59, 68
  density, 6
  elasticity, 57, 58
  in evolutionary research, 16
  formation, 65
  igneous, 35, 58
  magnetism, 5, 35, 43, 55, 72
  metamorphic rocks, 65, 69, 232
  oldest recorded, 50
  Precambrian, 65
  radioactivity in, 20
  reading list, 92, 96, 98

Index

structures, 4
Web sites, 102, 106, 118
see also Geology; Minerals; Mineralogy; Petrology
Roosevelt, Theodore, 31, 32
Rose, Susanna van, *Earthquakes*, 97
Rowland, F Sherwood, 47, 62
Royal Geographical Society Web site, 115
Rubey, William, 41
Rubidium-strontium dating, 42
Rudwick, Martin *The Meaning of Fossils*, 96
Rutherford, Ernest, 10, 59

San Andreas Fault and Bay Area Web site, 115
San Francisco earthquake (1906), 24
Satellites, 50
  Discoverer 1, 41
  ERS remote sensing, 52, 53
  GEOS 1, 44
  Geostationary Operational Environmental Satellite (GOES), 47
  geosynchronous orbit, 40
  Lageos (Laser Geodynamic Satellite), 48, 49
  Landsat, 46, 47
  Magsat, 49
  remote sensing devices, 26
  Seasat 1, 48
  Sputnik 3, 41
  Telstar, 40
  Tiros, 42, 43, 50
  Vanguard, 41
  see also Space
Schuler, Max, 31
Science Odyssey: You Try It: Plate Tectonics Web site, 115
Scotland, geological structure, 58
Scott Polar Institute, 53
Scripps Institute of Oceanography, 44
Sea *see* Ocean; Oceanography
Seafloor spreading, 5, 42–4, 62, 65, 68–9
  and hot springs, 19
  JOIDES project, 44, 46, 50
  ocean floor drilling, 44, 46, 47–50
  proof of, 43, 46
  theory, 21–2
  see also Continental drift; Plate tectonics

Sealife: A Complete Guide to the Marine Environment (Waller), 97
Seasat 1 satellite, 48
SEDCO/BP 471 drilling ship, 50
Sedimentary Environments: Processes, Facies and Stratigraphy (Dasch), 92
Seismic oil prospecting, 34
Seismic Tomography (Nolet), 95
Seismic waves, 63, 72–3, 74
  compressional waves (P waves), 6
  oil exploration, 5
  reading list, 95
  research, 68
  scientific uses, 31, 37
  transverse waves (S waves), 6
  velocities, 6–7, 9, 42
Seismograph, 58
Seismological Society of America, 87
Seismology, 63
  Earth core theories, 6
  earthquake prediction, 23, 24, 48
  explosion seismology, 35
  faults, 45
  global seismic tomography, 6
  instruments, 58
  Large Aperture Seismic Array, 44
  measurement of Earth's crust, 65
  organizations, 87
  recording earthquakes, 5–6
  Richter scale, 74, 231
  theoretical, 57
SEPM (Society for Sedimentary Geology), 88
Shoemaker, Gene, 27
Siebert, Lee *Volcanoes of the World: A Regional Directory, Gazetteer and Chronology of Volcanism During the Last 10,000 Years*, 96
Sigurdsson, H
  Encyclopedia of Volcanoes, 96
  Melting the Earth: the History of Ideas on Volcanic Eruption, 96
Silicates, 59
Silicon-oxygen bond, 4
Simkin, Tom *Volcanoes of the World: A Regional Directory, Gazetteer and Chronology of Volcanism During the Last 10,000 Years* 96
Siver, Raymond *Understanding Earth*, 96

Index

Skinner, B J *Blue Planet: An Introduction to Earth System Science*, 96
Smith, James *Introduction to Geodesy: the History and Concepts of Modern Geodesy*, 96
Society for Organic Petrology, 88
Society for Sedimentary Geology (SEPM), 88
Soddy, Frederick, 33, 59
Soil pH — What It Means Web site, 115
Soil science, organizations, 89
Soil Science Society of America, 89
Solar Energy: Basic Facts Web site, 116
Solar physics, 65
Solar System, 11
    geology, 16, 27–29
    research, 27–9
Some Like it Hot Web site, 116
Sonic depth finder, 33
Sound waves, sonic depth finder, 33
Space
    exploration, 18, 28–9
    research, 41–2, 46–9
    see also NASA (National Aeronautics and Space Administration); Planets; Satellites
SpaceCom Web site, 116
Sphinx Rock meteorological station, 37
Sprites, 53
Sputnik 3 satellite, 41
Stanley, S M *Earth and Life Through Time*, 97
State of the Climate: a Time for Action Web site, 116
Stommel, Henry M(elson), scientific contributions, 74–5
Storm Chaser Home Page Web site, 116
Storms, 43
The Story of the Earth (Dunning et al), 93
Stratigraphy, 66
    reading list, 93
Stratosphere, 31, 68
Streamflow and Fluvial Processes Web site, 116
Stripes of magnetic polarity, 22
Subduction, 22
Submarines, *Triton*, 42
Submersibles,
    ALVIN, 19, 48 *see also A Water Baby* (Kaharl)

Summerfield, M, A *Geomorphology and Global Tectonics*, 97
Global Geomorphology: an Introduction to the Study of Landforms, 97
Supercontinent, 20, 32, 63
Suspect terranes, 23, 49
Swinnerton, H H *Fossils*, 97
Sykes, Lynn R, 45
Synthetic diamonds, 7, 39, 60
Synthetic materials, 60

Tassy, Pascal *The Message of Fossils*, 97
Tectonics, 62 *see also* Plate tectonics
Tectosphere, 22, 44
Teisserenc de Bort, Léon-Philippe, 31
Telstar communications satellite, 40
Temperature
    atmospheric, 31
    of Earth's climate, 13
    of Earth's interior, 10
    global fluctuations, 33
    and greenhouse effect, 51
    monitoring from space, 52
    record-breaking, 50
    scales, 43
    stratospheric, 68
Temporal Urban Mapping Web site, 117
Terranes, suspect, 23, 49
Tesla, Nikola, 31
Theory of the Earth (Anderson), 91
Thermal consequences of subduction, 46
Thermodynamics, 37
    chemical, 2–4
    of open systems, 41
    rock/fluid reactions, 3
    thermal consequences of subduction, 46
Thiemens, Mark, 12–13, 50
Third World aid, 55
This Dynamic Earth: The Story of Plate Tectonics Web site, 117
This Dynamic Planet Web site, 117
Thunderstorms and Tornadoes Web site, 117
Thurman, H V *Essentials of Oceanography*, 97
Tiros satellites, 42, 43, 50
To a Rocky Moon: A Geologist's History of Lunar Exploration (Wilhelms), 98

Index

Toit, Alexander Logie Du, scientific contributions, 63
Tomography, global seismic, 6
Tornado Project Online Web site, 118
Torrey Canyon, 45
Torsion balance, 5, 34
Transform faults, 22, 69
Triton submarine, 42
Tropical Savannas CRC: Landscape Processes Web site, 118
Tropical Weather and Hurricanes Web site, 118
Troposphere, 31
Trueman, M A *Geology and Scenery in England and Wales*, 97
Tsunami! Web site, 118
Tsunamis, since 1981, 237–9
Tube worms, 19, *19*, 48
Turcotte, Donald L, 46
Types of Rocks: Igneous, Metamorphic and Sedimentary Web site, 118

UK and Ireland Climate Index Web site, 118
Understanding Earth (Press and Siver), 96
UNEP (United Nations Environment Programme), 46, 51
Unexplored Spaces Web site, 119
Uniformitarianism, 28
United Nations (UN)
  Basel Convention, 53
  Conference on Environment and Development, 52
  Environment Programme (UNEP), 46, 51
  World Meteorological Organization, 56
Uranium-lead dating method, 42, 59, 69
Urban-Rural Population Distribution Web site, 119
Urey, Harold Clayton, scientific contributions, 11, 17, 38, *75*, 75–6
US Committee on Atmosphere and Biosphere, 49
US Environmental Protection Agency, 52, 53
US Geological Survey Web site, 119
US National Science Foundation, 44
US Navy, 41

USGS Earthshots: Satellite Images of Environmental Change Web site, 119
U-Shaped Valleys and Truncated Spurs Web site, 119

Van Allen belts, 41
Van Allen, James, 41
Vanguard satellites, 41
Van Rose, Susanna *Earthquakes*, 97
Venus, 29
Vesuvius, Italy Web site, 119
View of a Sustainable World Web site, 120
Vine, Frederick John, scientific contributions, 22, 43, 76–7
  Global Tectonics, 95
Virtual Cave Web site, 120
Virtual Fieldwork Web site, 120
Volcanic activity
  effects on Earth's atmosphere, 16
  eruptions 20th century, 240
  and extinction theory, 18
  Mt Pelée eruption, 23
  prediction, 24
  reading list, 91, 92
  research methods, 24
  undersea hot springs, 19
  Web sites, 99, 112, 119, 120
Volcanic Successions, Modern and Ancient (Cas), 92
Volcanoes
  and Earth's atmosphere, 16
  monitoring, 24
  reading list, 93, 96
  Web site, 120
Volcanoes: A Planetary Perspective (Francis), 93
Volcanoes of the World: A Regional Directory, Gazetteer and Chronology of Volcanism During the Last 10,000 Years (Simkin and Siebert), 96
VolcanoWorld Web site, 120
Von Neumann, John, 38
Von Weizsacker, Carl Friedrich, 38
Vostok Lake, 53–4
Vostok station, 50

Walcott, Charles D, 32
Waller, Geoffrey *Sealife: A Complete Guide to the Marine Environment*, 97

Walsh, Don, 42
A Water Baby (Kaharl), 94 *see also ALVIN* submersible
The Watershed Game Web site, 117
Waves, Tides and Shallow Water Processes (Oxford University Oceanography course team), 95
Weather
  balloons, 37
  cloud-seeding, 38, 40
  cyclones, 39
  El Niño effects, 49, 55
  fronts, 58–9
  hurricanes, 40
  hurricane-seeding, 38, 43
  organizations, 86
  patterns, 61–2
  and pollution, 51
  reading list, 91, 92, 93
  satellites, 41, 42
  storms, 43, 53
  temperature scales, 43
  weather stations, 37, 39, 58
  Web sites, 103, 104, 107, 109, 112, 114, 116, 117, 118, 120
  see also Climate; Meteorology
The Weather Book (Hardy, Wright, Gribbin and Kington), 94
Weathering, 241
The Weather Machine (Calder), 92
Weather: What forces affect our weather? Web site, 120
Wegener, Alfred Lothar, scientific contributions, 20, 32, 69, 77–8
Wegener's hypothesis, 77, 77
Weizsacker, Carl Friedrich von, 38
Welcome to Coral Forest Web site, 121

Wells, Sue *The Greenpeace Book of Coral Reefs*, 98
Wetherald, Richard T, 45
Whaling, 55
Whittow, John *Disasters*, 98
Wilde, Simon, 50
Wilhelms *To a Rocky Moon: A Geologist's History of Lunar Exploration*, D, 98
Wilson, Edward, O *The Diversity of Life*, 98
Wilson, John Tuzo, scientific contributions, 22, 44, 78
Wilson, M *Igneous Petrogenesis, a Global Tectonic Approach*, 98
Windley, B F *Evolving Continents*, 98
WOCE (World Ocean Experiment), 52
Woodhouse, J H, 6
Woods Hole Oceanographic Institution, 35
Woods Hole Oceanographic Institution Homepage Web site, 121
World Climate Research Programme, 48
World Environmental Changes Landmarks Web site, 121
World Meteorological Organization Web site, 121
World Ocean Experiment (WOCE), 52
World of Amber Web site, 121
Worldtime Web site, 123
Worldwide Earthquake Locator Web site, 123
Wright, Peter *The Weather Book* 94

X-ray crystallography, 67
X-rays, 3, 8

Zircon crystals, 50